Everyday AMAZING

Fascinating Facts about the Science That Surrounds Us

Beatrice the Biologist

Adams Media
New York London Toronto Sydney New Delhi

Adams Media
An Imprint of Simon & Schuster, Inc.
57 Littlefield Street
Avon, Massachusetts 02322

First Adams Media trade paperback edition May 2019

ADAMS MEDIA and colophon are trademarks of Simon & Schuster.

For information about special discounts for bulk purchases, please contact Simon & Schuster Special Sales at 1-866-506-1949 or business@simonandschuster.com.

The Simon & Schuster Speakers Bureau can bring authors to your live event. For more information or to book an event contact the Simon & Schuster Speakers Bureau at 1-866-248-3049 or visit our website at www.simonspeakers.com.

Interior design by Sylvia McArdle
Interior images by Katie McKissick

Manufactured in the United States of America

10 9 8 7 6 5 4 3 2 1

Library of Congress Cataloging-in-Publication Data
Names: McKissick, Katie, author.
Title: Everyday amazing / Beatrice the Biologist.
Description: Avon, Massachusetts: Adams Media, 2019.
Includes index.
Identifiers: LCCN 2018061006 | ISBN 9781721400287 (pb) | ISBN 9781721400294 (ebook)
Subjects: LCSH: Science--Popular works.
Classification: LCC Q162 .M455 2019 | DDC 500--dc23
LC record available at https://lccn.loc.gov/2018061006

ISBN 978-1-72140-028-7
ISBN 978-1-72140-029-4 (ebook)

Dedication

For Fiona—may you be as easily amused as your mother.

Acknowledgments

A big bucket of thanks goes to my supersmart friends who indulged several dumb questions from me while writing this book. Jessica Parr, Aaron Celestian, Jann Vendetti, Robert McNees, Libby Ellwood, Andrew Williams, Dean Pentcheff, Assal Habibi, and Scott Perl—thank you for your brains and letting me pick them.

Contents

Introduction

Every day I do a lot of amazing things. I make a cup of tea, warming water up to the point where the molecules are so energetic they want to leave the confines of their liquid form to assume steam, at which point I add dried leaves that bend to the will of this hot water and release a stream of chemicals including much-anticipated caffeine. I drive a car, propelled forward thanks to a controlled explosion of energy stored by marine plankton millions of years ago. At the grocery store I buy produce that built its stems and leaves and fruit by using the light of a star burning 93 million miles away.

Meanwhile, I constantly breathe in the surrounding air, extracting oxygen molecules and passing them into my bloodstream, where they piggyback on red blood cells for delivery to my many different types of cells, including ones in my brain that use the oxygen to keep the cellular machinery running so that neurons can receive information from my optic nerve and piece together the images I see before me.

And this is an average day. Now, I don't want to lead you to believe I dissect each of my activities like this in real time. I simply don't have the bandwidth for that. For the sake of efficiency, we simply have to take some things for granted. However, it's nice to take a step back from the everyday routine to ponder the everyday amazing—the little things we experience every day that are quite spectacular.

So let's look at the more than sixty entries throughout this book that give you information on everything from what atoms are made of, how fast a radio wave just flew past you, and why you can't always trust your own brain. Let's take nothing for granted, examine the usually unexamined, and look at the often overlooked. Come on. It'll be fun.

CHAPTER 1

Teeny Tiny Building Blocks: Atoms and the Super Weird Stuff They Do

I bet you take atoms for granted. You probably expect them to just continue to make up the mass of your body, the chair you're sitting in, the liquid in your water bottle, and the air you're breathing. But hey, don't worry; we all do it. We forget about the teeny tiny building blocks that create the foundation of our world and the super weird stuff they do. It's a forgivable transgression. They're just annoyingly, obnoxiously small, and invisible things have a way of going unnoticed. So really, it's all their fault.

But try for a moment to think about what all the materials around you are made of. Your skin. The couch. Your dog. They're all made of little itty-bitty atoms. So are the things too small for you to see (dust mites, bacteria, viruses) and the things too big for us to fathom (faraway stars, dusty nebulas, galactic swirls). When you get down to it, everything is atoms. Us too. And you'll learn plenty about what atoms are and how they work here in this chapter. So let's get to it.

What We're *Really* Made Of: The Scoop on Atoms

Let's back up a minute. Admittedly, maybe for two minutes. Atoms are the units that make up the stuff that's all around us. They have a nucleus, a densely-packed center with a varying number of bits called protons and neutrons, which are preposterously small particles. Around the nuclei of these atoms is a swirling cloud of electrons—in some cases only one, and in other cases dozens and dozens.

Showing an atom to scale in a drawing is nearly impossible. If the nucleus is the size of the period at the end of this sentence, the electrons are swirling in a cloud around it up to 32 feet away, making the total width of this example atom 64 feet. I don't have that much paper.

That flurry of electrons is what establishes the border of an individual atom. And much of the way that atom behaves and interacts with other atoms is based on what those electrons do.

As I lightly tap the keys of my laptop, my skin comes in contact with each plastic square. I can feel as the atoms in my skin cells are mashed against the atoms in the plastic, coming to a point where they push up against one another. I think of it as a keystroke, but really it's an awkward atomic encounter. My atoms and my laptop's atoms never actually touch each other. Instead, they get to a point where the electrons swirling around in each of the atoms come close enough to be repelled by their negative charges, like what happens when two opposing magnets are pushed against each other.

That means we never, atomically speaking, come into contact with anything. It's just electrons repelling each other all day long. Even now, in between you and the chair you're sitting in or the ground you're standing on (for I don't know how you prefer to read your books) is a shred of empty space. You're kind of sort of floating. Doesn't it feel great?

Weirder still, the atoms that make up both the keys and my fingers (and everything else) are made mostly—almost entirely, even—of empty space. The "stuff" we're made of has surprisingly little actual stuff. It's hard to wrap your head around. Everything we perceive is so definitely space-filling. My hand, the wall, my fuzzy slippers. They've all convinced me that they're incredibly solid materials. But they're still made of these mostly empty atoms. I can't even call them "airy" atoms. Because air is yet more atoms.

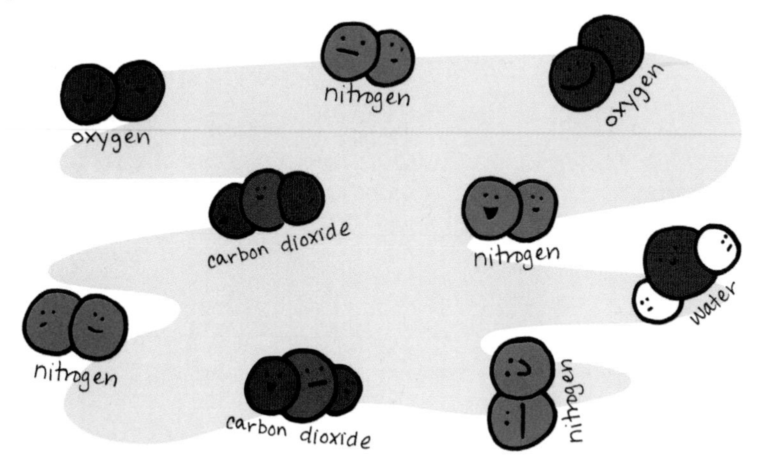

Odds are you're too busy on an average day to stop every ten seconds and awe at atoms and marvel at molecules (which are certain groupings of atoms), but I highly recommend a daily regimen of atomic reflection. There are big ideas wrapped up in these little packages. While they're so small we easily forget them, it's ultimately atoms' behavior that drives our experience of the world around us, and you literally wouldn't be anything without them.

So, Hey, Where'd You Get Your Atoms?: Why We Really Need to Eat

I know it's common (if somewhat cliché) to say "you are what you eat"—I mean, what the heck else would you be?—but it's amusing to further ponder this. Take a moment to consider your body, from your skin to your bones to the squishy brain in your skull. Where did all that *stuff* come from? Where, exactly, did you get all those atoms?

Your mom got you off to a good start with a few of them, but ever since then, you've been getting atoms from the things you eat and drink each day. It's easy to forget about this as you go through your routine. When you go out to lunch, you're probably thinking more about satisfying your hunger than about your need for atoms.

But don't feel bad. Atoms are so easy to skip over. You might be aware of food going through you (though mostly when you're cavorting with a toilet), and sometimes you can feel a snack's immediate effects—a caffeine buzz or sugar rush, for example. You can be satiated by a nice meal, but what exactly do you use all that food for? (Besides stopping your empty stomach and intestines from growling at you?)

The truth is that you need the atoms from your food and the energy stored in the bonds between those atoms. You need calcium in your diet to build your bones. You need sodium to send electrical impulses between neurons in your brain. You need iron to help your red blood cells carry oxygen. You need sugars, which are specific arrangements of atoms, for your cells to get the energy they need to do their many jobs.

When my daughter was a newborn baby and all she had ever ingested was breast milk, I would often marvel at the fact that every atom in her body had come from me (except for the oxygen atoms she inhaled since she drew her first breath of air). Her body was built from the food I ate when I was pregnant, as well as a few atoms she may have stolen from my body's own reserves, such as the calcium from my bones. And once out in the world, drinking breast milk multiple times a day, she was still taking atoms from me to build her body. Now she has the decency to get her atoms elsewhere, from applesauce and Cheerios and string cheese, but for a while there it was just me. I was a one-stop shop—Atoms 'R' Us.

Wherever you get your atoms, they are merely borrowed, and they never go to waste. The ones in my body today had lives before me, and they'll move on after I die. They used to belong to plants or animals (which I ate), and before that, they were in the air, the soil, and in previous generations of living things, going back to the very beginning of life on this planet. A carbon atom in one of my skin cells could once have been part of a dinosaur. An oxygen atom in my liver could once have belonged to a trilobite.

Welcome to the neighborhood! How did you get here? I'm always curious.

Oh, hi! I was in a grain of rice. My fondest memory, though, is the time I was part of a dinosaur.

That's fantastic! I remember being poop a while back.

Oh, yeah. Been there.

And before the first cell graced our planet with its presence, those atoms were the components of a young Earth, and earlier still, they were space dust swirling through an early solar system, which itself was forged in the bellies of giant stars many billions of years ago that exploded and sent atoms out into the universe. And now I just use them to peruse social media.

When I'm done using my atoms (hopefully in, like, sixty years or so), they will continue to be awesome, possibly far more so than they ever were with me. The carbon in my body can be taken by bacteria who will use it to make more bacteria, which will be eaten by a worm, which will be eaten by a lizard, which will be eaten by a falcon. But these creatures, too, will merely be borrowing the carbon for a limited time, as those atoms ultimately belong to the universe.

Getting Salty: Meet the Everyday Crystal, Sodium Chloride

I've gotten you thinking about atoms a bit, but now we're going to level up and think about chemicals. Does that word make you squirm a little? The word *chemical* has somewhat of a negative connotation surrounding it these days because it's often used in place of a more accurate word like *toxin*. Chemical is a very general term, and being called a chemical doesn't mean a thing is harmful. After all, think of *dihy-*

drogen monoxide. It sounds intimidating, doesn't it? But don't worry, in moderate amounts, dihydrogen monoxide (water!) has been shown to not only be safe, but actually quite necessary for your health.

Another chemical (now that word doesn't seem so scary, eh?) we're quite familiar with is sodium chloride, or salt. There are many, many different types of salt, but this one is among the most common, and it's the one we sprinkle over our food since our bodies perceive it as quite the flavor enhancer.

But what is salt? Have you taken a close look at it lately before it cascades onto a piece of avocado toast?

It's a crystal. Ooooooh.

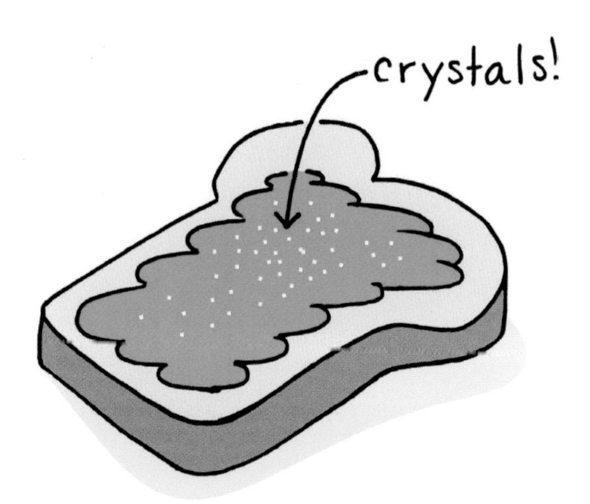

Crystal is a technical term describing a material with a highly ordered, repeating chemical arrangement—which often makes it ridiculously good-looking. Diamonds, rubies, and emeralds are crystals, but they don't own the rights to crystallization. Ice (like, frozen water) is a crystal, as is sand (which is often made mostly of quartz), and as I previously mentioned, salt.

But what *is* salt? The salt in our food and the salt in the ocean are both mostly sodium chloride. But what does that even mean?

Let's first talk about sodium for a second. Sodium is an element that, in its pure form, is a shiny metal, which looks absolutely nothing like table salt. It also doesn't behave anything like it. If you drop a bit of pure sodium into a glass of water, it will cause a small explosion. I don't have too many qualms telling you this because you probably can't get your hands on pure sodium without having a few moments to rethink the terrible plan you're hatching. I would urge you to watch a video of this phenomenon instead.

The sodium atoms in salt are a little different. They're sodium *ions*. Each of the sodium atoms in your salt is missing one electron from the swirling cloud around its nucleus. It turns out that the loss of just one electron is all it takes for a sodium atom to change from part of a highly reactive metal to a component of very tasty salt.

Where did that electron go, you might wonder? In the case of table salt, it was lent to an atom like chlorine, which with the addition of an electron becomes an ion as well, which we call chlor*ide* (the *-ide* is meant to let you know it's an ion). And just like sodium, this ion acts completely differently than its elemental form, which is chlorine gas—famously used in World War I to, uh, kill people.

Truly, it is beyond bizarre that a highly reactive metal and a poisonous gas, when put together and an electron is swapped between each pair of these atoms, can become a taste-enhancing crystal. And we haven't even begun to talk about the party trick salt can do. When you put it in water, it disappears. Bizarre! Again, this is one of those things we see all the time and don't have the bandwidth to talk about. For instance, I made pasta last night, and I added salt to the pot of water into which I would eventually throw many chunks of glutenacious swirls. And then, soon after, no salt crystals remained. Where did they go? They were dissolved into the water. I mean, wow. I like to sit there while I make pasta and think about the salt crystals slowly being pulled apart by water—yup, that's what's happening. Because water is a molecule with a positive end (the side with the hydrogen atoms) and a negative end (where the oxygen is), it can interact with both the positively charged sodium ions and the negatively charged chloride ions. The water finds these ions hopelessly attractive, and they surround them like eager fans around a sexy celebrity (except more like zombie fans since they start to tear apart the celebrity in this analogy... wow, this took a dark turn). As more and more water molecules do this, the salt slowly gets pulled apart until all the individual ions are surrounded by water molecules, and to our eye it has completely disappeared.

What's even cooler is that if all the water evaporates away, the salt gets left behind and turns back into a crystal. If you have sea salt in your cupboard, that's how it was produced. The manufacturer got salty seawater, banished all the water, and what's in your cabinet is the salt that was abandoned.

While regular tap water doesn't have nearly as much salt as seawater does, it still has dissolved ions floating in it, which we can tell because of the hard water crust left at the bottom of the dishwasher or clogging the tiny holes in the shower head. Don't be annoyed by them. They're just crystals that got left behind when the water evaporated away. They've been through a lot, actually. And they could use a friend.

So take a moment today to marvel at the crystals around you—whether it's the salt you add to a dish you're cooking, or the crusty stuff in your shower.

Hot and Cold: The Fabulous Physics in a Cup of Coffee

Apart from promoting the splendor of salt crystals, I'm here to convince you that even something as simple as a cup of coffee slowly cooling off is quite exciting. Don't believe me? Let's discuss.

Even in the summer, I prefer drinking hot coffee. Call me whatever you want, but I simply don't like the taste of coffee when it's cold. But what's so darn different about hot coffee, cold coffee, or (shudder) room temperature coffee? It's still coffee, after all. Why would people prefer it a certain temperature?

Surprisingly, what we're discussing is how much we want the atoms in our coffee to vibrate—that is, how much energy they would ideally have. I want the atoms in my cup of coffee to be vibrating as fast as I can handle it, just not so much that they burn my tongue.

Temperature, coffee related or not, is a funny thing. We think about it a lot, whether it's the air temperature outside and inside (especially if you have the air conditioner on), our body temperature (especially if you are running a fever), and our food temperature (especially if you are cooking meat). But we don't normally reflect on what temperature is actually measuring or what that number means for the substance we're concerned about.

Temperature, basically, is a way for us to describe how much heat there is in something. But, then, what is heat? (I know, I know. I feel like a toddler constantly asking "why?" over and over. But stay with me!) Heat is energy. And an interesting thing about heat is that it doesn't sit still. It's always on the move, going from something hotter to something colder.

The heat in my hot coffee right now is slowly leaving. In fact, the steam I can see rising from my cup right this very instant is a gorgeous display of chemistry and physics I can admire between sips. The water molecules in my coffee are so excited and full of energy they aren't content to stay bunched together as liquid. They literally can't contain themselves so they start to float away as water vapor. Have you ever been so excited about something you couldn't help but jump for joy? Yeah, neither can the water in my coffee mug.

Over time, all this heat loss makes my coffee cool down. The heat still exists; it's just somewhere else, like the surrounding air, or in the warmth of my hand curled around the cup. There is a rule in physics about energy. In fact, this rule is seemingly so ironclad we call it a law. It states that energy (and matter, for that matter) cannot be created or destroyed. It doesn't whoosh in and out of existence like some magical ether; it only moves around.

So even when you do something as rote as blow on a hot cup of coffee you're running a quick physics demonstration. You science wizard, you—you're making the transfer of heat more efficient. The air you're blowing allows the coffee to interact with more molecules, which can grab a bit of the heat away as they whoosh past. When you sit in front of a fan on a hot day, you're doing the same thing.

When you get down to it, whether you're a person, a cup of coffee, or an individual water molecule, we're all merely a means for heat to move around, which is a great thing to ponder during a Sunday afternoon existential crisis. I speak from experience.

The War That's Waged When We Wash Our Hands: How Soap Actually Works

Have you washed your hands today? Unless you are reading this mere moments after waking up, I really hope you have. Otherwise, we can't be friends because I am quite the germophobe (technically the fear is called *mysophobia*).

Soap will be the item I miss most in our increasingly likely post-apocalyptic future. I take comfort in having clean hands. But why do I care so much?

Even an avid handwasher like myself can forget what I'm achieving when I lather up and rub my hands together for a while in the sink. But, if my palms are especially greasy, it becomes more obvious why water alone is simply not going to cut it, so it's easier to admire what soap is doing for me.

Many of the things we're trying to remove from our hands are oily in one way or another, whether they're olive oil, machine oil, dirt, or perhaps a bit of bacteria. Water does very little to remove these things. If you want to test this, rub butter all over your hands and then head to the nearest sink. Even with running water dousing your hands, that butter's not going anywhere, is it? (Please note that this experiment is a tragic waste of perfectly good butter.)

But there's a perfectly good explanation for this. The water coming out of your faucet, as I said earlier (and which we'll talk more about later), is an interesting molecule because it's *polar*. This means that, while overall it's neutral, it still has a positive side and a negative side. As my grandfather used to relish saying when the opportunity arose, "I know some people like that." Oil, on the other hand, is *nonpolar*—molecules in oil don't have positive or negative ends. It's quite neutral all over.

And when it comes to charges (positives and negatives), you start dealing with attractions and repulsions. Two negatives repel each other, but also, a polar molecule like water doesn't interact much with a totally neutral, nonpolar substance like oil.

Oil and water's reluctance to mix is relatively well known. We'll say two people don't get along by noting they're like oil and water. They don't explode or anything, thank goodness, but try as you might, you can't get oil to dissolve into water the way that something like salt will. The term for things like oil is that they're *hydrophobic*—afraid of water. For that reason, when you run water over your hands, it does very little to remove the oils, and any of the microbes hiding out there as well.

So what is so special about soap that it's able to mediate this oil and water standoff? Soap brokers peace between oil and water because it has both an oil-loving side and a water-loving side. The technical term for this is *amphipathic,* which to an untrained ear might make it sound like soap can read minds or predict the future, but that's not it.

What this means basically is that when you lather soap all over your hands, the oil-loving side of the soap molecules gets friendly with the grease on your hands. They mingle and discuss their days. Once they've all settled together, you, dear hand-washer, run water over them. Now, the water-loving side of the soap molecules get swept up, taking the oil-loving side and all the grime it has fraternized with off of your hand, and into the bottom of the sink, going down the drain and far away. Yet, just like energy is neither created not destroyed, the same goes for filth. You haven't truly gotten rid of any dirt; you've simply moved it someplace else, in this case the sewers.

The oil was happy to ignore water, but it was tricked into going along with it by soap. Soap is quite the double agent in that sense. It plays both sides, which I'm happy about because it means I can get the gunk off my hands.

Hopefully the next time you wash your hands (or are tempted to believe that water alone will do), you can think about what is happening when you lather, rub, and rinse. Soap isn't a mystical substance that cleans your hands. It's a chemical link between oil and water that otherwise wouldn't jibe. You're washing your hands with the fundamentals of chemistry.

And if you're the lady I saw in a public bathroom the other day who didn't wash her hands at all, stop being gross.

Atoms in the Oven: The Chemistry of Toasting Bread

I find that people often think chemistry is only done in expensive labs with brightly colored, toxic concoctions, but really, every time you cook you're conducting a chemistry experiment. Even seemingly simple tasks like making toast are actually bajillions of chemical reactions in disguise.

Now, why *do* we put bread in the toaster with such frequency? We could eat it plain. And if you insist on eating it warm, you could microwave it maybe. But there is something different about putting a slice of bread in a hot toaster. Why is that?

Even though no one ever talks about it, we consider this kitchen appliance a must-have because, at the high temperature in the toasty toaster, we transform boring, bland bread into magical, delicious toast. It's not just warm bread. There's so much more going on than a change in temperature.

At those high temperatures you kick-start certain chemical changes called Maillard reactions. These happen between amino acids, which are the building blocks of proteins and sugars. These chemical reactions make completely new compounds, many of which we perceive as quite delicious. In bread, toasting yields flavor compounds that taste like caramel. Mmmm.

There are loads of foods that taste oh-so-much better when they're toasted or roasted, whether it's coffee beans, popcorn, meat, or countless others. Heat changes them and creates brand-new flavors that wouldn't be there otherwise. You're not just warming your food; you're creating new things!

This is also why so many foods taste completely different if you roast them versus boil them. Boiling water simply can't get hot enough to make those reactions happen. No matter how you slice it (pun definitely intended), boiled chicken (or carrots for my vegetarian friends) will never taste as delectable as oven-roasted chicken (or carrots).

But of course it's possible to overdo it. Burnt toast has been heated enough to make yet more new compounds, some of which are bitter.

So don't you dare tell me you're not a chemist and that whole subject is alien to you. As long as you can make toast, you're an accomplished scientist and a pretty impressive chef, I might add.

Most Mysterious: The Deal with Dark Matter

There's one more thing. I don't quite know how to tell you this, but I'll try. You, right at this very moment, are surrounded by a form of matter no one fully understands. Reach your hand out and wave it around a bit. If you're in public, pretend there's a fly you need to swat away or something. Or do some jazz hands and weird out any random passersby. There, now you've just kind of, sort of touched some dark matter. Except you didn't because our bodies can't detect it. No one can. Not even with the fanciest equipment. You can't see it. You can't feel it. Can't smell it. Can't hear it. Can't have dinner with it.

When describing dark matter, it starts to sound a lot like a ghost. It's something somewhat familiar, but it exists in a different way, one that makes it currently impossible to measure or observe, but we still know it's there. (Except dark matter is far more likely to exist than ghosts.)

We know dark matter is out there because its presence can be measured in the behavior of huge galaxies. The "regular" matter alone doesn't account for how things are moving way out there. There has to be more "stuff" for all the math to work out. Dark matter is one of those things that only seems to matter when there is lots of it—like how much is around an entire galaxy, such as our Milky Way. It's harder to see up close, which is why even though it must be everywhere, we can't seem to find any of this dark matter.

And still, we're swimming in it. It passes through us every moment of every day. We have been blissfully ignoring it, and once you finish this chapter, you can resume that state. Just like we forget that our bodies are made of atoms that are mostly empty space, we forget about the dark matter that makes up so much of our entire universe.

CHAPTER 2

The Waves All Around Us: A Tour of the Electromagnetic Spectrum

Now, don't freak out, but at this very moment, no matter where you are, you're surrounded by electromagnetic radiation. Before you go running for a tin foil hat, gas mask, or a bed to hide under, allow me to let you in on a little secret—this radiation is normal (and awesome).

Radiation sounds so scary, and some of it definitely is, but a lot of it is perfectly harmless and, in many cases, useful. If you could please reach your hand out again, try to feel a passing radio wave (when you're done trying to detect dark matter). You can't feel it? Yeah, me neither. But it's there. I know because if I turn on a radio, voices come out.

In this chapter, we'll explore the many waves that surround us, like the waves we use to see, cook our food, and listen to the dawn of the universe.

Welcome to the Universe!: You're Surrounded by Waves

In the previous chapter, we talked about matter—actual, tangible stuff—along with the very intangible dark matter. But that's only part of our everyday story. There are also a lot of energy waves, and they do just as much to make up our world. Some of these waves we can detect, and others pass by us (or even through us) without us knowing it. They probably would like to say hello as they go by, but our bodies don't have the means to sense most of them.

These energy waves can be any length, and to better understand them, we line them up in order on a spectrum and give them fancy names. This lineup is something we named the electromagnetic spectrum, which may sound incredibly intimidating but it's not so scary if we tease it apart. It's called "electro" and "magnetic" because all of the waves travel through the universe as disturbances in electric fields and magnetic fields. But we don't have to get into all the physics of that to appreciate them.

The longest waves on the electromagnetic spectrum are called radio waves, and as they get shorter, we start calling them microwaves. Next, there are infrared waves. The very narrow portion of waves the human eyeball can detect we've dubbed, in our usual anthropocentric way, "the visible spectrum." The waves just a little too short for us to see are called ultraviolet rays. And then we have X-rays, which are even shorter than ultraviolet, and the shortest we'll discuss here are gamma rays.

You may have picked up on the fact that as waves get shorter, they start to sound more dangerous. And that's true. The shorter a wave is, the more energy it carries. We also say they have a higher frequency, which is how many times the wave goes up and down in one second. It would be easier to say that it goes "faster," but that's not quite right. In my opinion, one of the weirdest things about these different waves is that they all travel at the same speed—no matter if they're a radio wave, visible light, or an X-ray. You may have heard of this speed, even: it's the speed of light.

While you're sitting there reading, waves are whizzing past you at light speed, which is 186,282 miles every second. Now, we all kind of know that light is fast; in movies, it's always very impressive if something can move at light speed. But can we try to visualize how fast that is? One hundred eighty-six thousand two hundred eighty-two miles is 7.5 times the circumference of Earth. That means if you could travel at the speed of light you'd whoosh seven and a half times around the entire planet in *one second*. That's absurdly fast.

Light speed in its sci-fi usage is nearly synonymous with "instantaneous," but let me tell you, out in aptly named space (where there is indeed no shortage of space) even light's dizzying pace can seem slow. When a visible light wave leaves the sun, it takes eight minutes and twenty seconds to reach us here on Earth. That means if you got into a hypothetical spaceship with a light speed function and set yourself on a course for the sun, you could listen to "Stairway to Heaven" in its eight-minute entirety and still have enough time left over to sing "The ABC Song." Right after belting out, "Next time won't you sing with me?" you'd go barreling into the sun, so it would be a pretty ironic way to go.

Let me reiterate that we can talk about all of these waves in terms of light speed. Radio waves, microwaves, infrared waves, ultraviolet waves (okay I'll stop listing now)—these are all moving at this speed. You might not think of them all as "light," but they are. Light is energy. A radio wave is an energy wave the same way that visible light is. We'll get into more details about these waves soon, but for now, just remember that there are waves all over the place, and they come in different sizes. And yes, they do sound quite exotic, but many of these waves are a part of our everyday lives.

You Hear the Voices, Too, Right?: Radio Waves

When you turn on your car radio, do you perchance hear music and voices? Do you question your sanity when this happens? Probably not because we take the process for granted at this point. But think about how the radio waves your car detects and translates into sound waves for you to hear are also floating by you all the time. Since you're not a radio, you can't really do anything with them, but they're still there.

And for a very reasonable price, you can even buy a device that generates its own radio waves, such as a walkie-talkie, baby monitor, or garage door opener. In fact, just about every fancy gizmo we use these days employs some radio waves in one way or another, like your cell phone and your wireless router.

But with all these radio waves flying around, sometimes you have to be aware of the waves your devices are emitting because they could conflict with someone else's.

That happened to me once as a friend and I conversed via a pair of $15 walkie-talkies while we drove up California's I-5 freeway in separate cars. Those handy little gadgets were set to transmit and receive the same length of radio waves, so I could talk to her over the varying 10- to 100-foot distance between us as we drove through central Californian farmland, changing lanes to go around giant big rigs full of tomatoes, garlic, and oranges. As thrilling as it was to generate, transmit, and receive each other's radio waves, the most notable part of this experience happened about halfway through our journey, when an unfamiliar voice came through my radio. "Get off this channel!" it said through some static.

"Uh...what?"

"Get off channel five!" the mysterious voice yelled.

With all the context clues available to me, I surmised that we were sharing this frequency with one of the farms we were zooming past.

Ensconced within my cloak of anonymity and the shield of a car going 75 miles an hour, I didn't hesitate to dissent. "You can't tell me what to do!" I said gleefully, adding, "You're not lord of channel five!"

And before the angry farmer could reply, we were quickly out of range.

I admit I hadn't thought about the exact frequency of waves my walkie-talkie was sending, but if you use radios more often, you have to be more conscientious. There is definitely an etiquette to making waves.

And let me mention that these aforementioned techy wonders are just the radio waves that we humans generate for ourselves on purpose. There are also radio waves from natural sources, and we can hear those too.

When you turn on your radio and browse through the options, you'll pick up signals from your local radio stations that generate waves on specific channels. They blanket an area with these waves so anyone in their car or with their clock radio can pick them up. But if you tune your radio to a channel no one is using, you get that familiar static white noise.

What is that snowy sound? Your radio is picking up on radio waves—if it weren't, it would be totally silent—but they're not ones that any station is sending out encoded with music and voices. It's just...noise. This assortment of radio waves comes from a combination of natural sources and feedback from other gadgets we use.

Some of the static comes from waves produced by Earth itself. So if you want to immerse yourself in the wisdom of Mother Earth, just listen to static in the car instead of NPR. But a small fraction of that static is from *outer space,* and it's called cosmic background radiation. And even though you can hear these waves with your *radio*—they're technically *micro*waves (which we'll get to in the following section) from every corner of space, which were first produced at the birth of the known universe 13.8 billion years ago. Whoa.

And once again we have to remind ourselves that they are around us all the time, these waves. We just can't tell (at least not without special equipment). They were there when our ancestors were single-celled critters floating in an ancient ocean. They were there, waving over the Earth as dinosaurs lumbered through prehistoric forests. And they're here still, only now we're clever enough to find them.

Not Just for Popcorn: Microwaves

But enough about the dawn of the universe; let's talk about other microwaves. And by the way, how cool is it that when I say "microwave," which literally means a "rather small electromagnetic wave," you probably picture a kitchen appliance? I'm glad they named the cooking gadget after electromagnetic energy so it could familiarize countless people with this form of light wave. They could have just as easily called it something silly like "cooker-looker," "very presto oven," or "popcorn cube."

Before we travel too far with microwaves (remember, they go at the speed of light, like all these waves do), I should mention there is no absolute, clear-cut line separating radio waves and microwaves. Remember, you can hear some microwaves with your radio. And by some definitions, microwave ovens use *radio* waves to cook your food. So on second thought, maybe we shouldn't name devices after waves because sometimes it's misleading.

But that's okay. It's a reminder that all the waves on this vast spectrum aren't here to follow our rules. Nothing is. The universe doesn't exist for us to understand it. We try to divide things up and assign them names and descriptions, but everything is messy. (Just ask "dwarf planet," Pluto.)

Getting back to waves that are micro, the eponymous kitchen appliance generates waves that transfer their energy into the food you put in there. The particular lengths of the waves these ovens use are especially good at exciting water molecules as well as fats. The transfer of energy heats up water and fat in our food and makes those molecules vibrate faster. This is not absurdly different from other forms of cooking; it's just much more efficient, which is why microwaves are well known for the speed at which they heat things up.

But as marvelous as microwave ovens are, they don't compare to the microwaves that surround us all the time, reaching here from outer space, where they've traveled gajillions of miles over billions of years, whispering cosmic secrets to us we cannot hear without sensitive technology.

Or maybe microwave popcorn is more notable. It's up for debate.

Smokin' Hot: Infrared Waves

When you make some microwave popcorn, the hot bag lets off yet another kind of electromagnetic wave: infrared radiation (but don't worry, it will be okay).

Infrared waves are a tad longer than the red section of visible light. Infrared means, basically, *below red.* We can't see infrared waves, but we use them every time we adjust the volume on our TV with the remote, and they're how fancy night-vision goggles work.

When there are sufficient infrared waves, we can actually feel them as heat. But they're still light waves just like all the others. We associate them with heat the same way we associate visible light with being, you know, visible. These waves happen to fall within the limits of what our bodies can detect when there are enough of them around. And while you can think of infrared radiation as heat, you don't have to be what we would consider "hot" to emit them. Even an ice cube emits some infrared radiation, just not nearly as much as a pizza fresh out of a brick oven. Technically, everything that is above absolute zero (the theoretical temperature so cold that atoms stop moving) gives off some amount of infrared radiation. It's everywhere.

Infrared waves are the only waves you generate on your own all day long. You don't produce any visible light; you only reflect it from other sources. You don't make radio waves. (I mean, I'd hope you would tell me if you were doing that. If you are, we need to talk.) But you make infrared waves all the time because you have body heat. You're glowing in a way that we can't see with our naked eyes.

I like that we tell people who seem happy and healthy that they're "glowing," but you can feel free to tell absolutely anyone this because it's always true, regardless of the state of a person's health. Even when I'm stuck in bed with the flu, battling depression, and reading the latest news, I am positively radiant, glowing with infrared radiation. You don't see it ordinarily, but with the help of night-vision goggles or an infrared camera you can see just how brilliant everything around us truly is.

Step out into the sun and take a moment to feel that heat. What you're feeling is infrared radiation that traveled all the way from the sun to get to you. It had to go so far. It cruised alongside the visible light that is now scattering through the atmosphere so you can see the blue sky and the ground at your feet. But the infrared is there too. These waves are what has made our planet habitable to begin with. If by some strange quirk our sun only emitted visible light, but no infrared radiation, we wouldn't be here to figure out how in the heck it's doing that.

And even when the sun isn't shining, we're getting infrared waves from all over the universe. It's weird to think when you look up at the night sky you're only detecting a small fraction of the energy out there. What we see of the stars and planets, just like everything we behold, are only the wavelengths of light that are visible to our eyes. There is also invisible infrared radiation reaching us from those places; our bodies just can't detect it (although special telescopes can). We experience but a fraction of the energy in the universe, and even that is totally overwhelming for me most of the time.

See for Yourself: The Visible Spectrum

After thinking about so many waves that we blissfully ignore each day, we can finally get to some that we can detect with our human eyeballs. Visible wavelengths of light carry enough energy to excite molecules in cells in the back of our eyes without damaging them—that's what makes them "visible." Well, to us, at least.

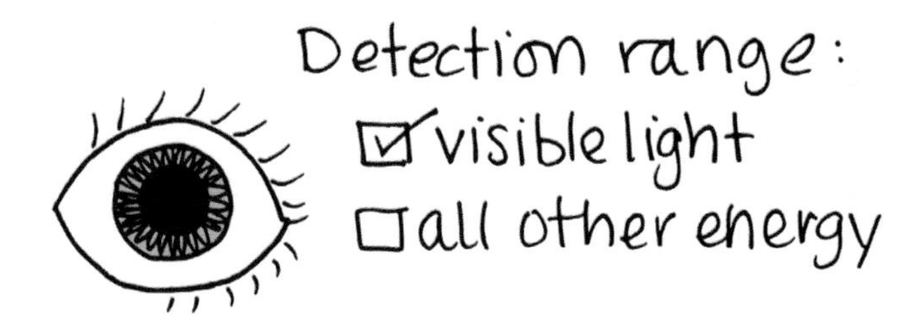

One of visible light waves' best features is that they bounce off plenty of materials. Whereas X-rays go right on through a lot of stuff, light by and large bounces around, and eventually, some of it bounces right into our eyeholes. Light can go through some things, of course, like glass (and we'll talk about that in Chapter 4).

All this bouncing is where things get fun. When I look at something, like a book sitting on my desk, I'm not seeing it as it truly is; I'm seeing the light that happened to reflect off it a fraction of a second ago. The same goes for colors. My red pencil is only red because the molecules that comprise it reflect red light toward my eye and absorb the other colors of visible light.

In total darkness, there is no color. I know it sounds terribly philosophical, but color is truly in the eye of the beholder. If there is no light to bounce off something, and no eye to sense that light, then there is no way for color to exist. Put that in your "stuff to think about when I'm sad" folder.

We can see what we see only because the structures in our eyes can differentiate between these wavelengths. Some animals like worms and scallops can see only light and dark; color doesn't exist to them. And even among humans, with heritable traits like color blindness (and there are many different kinds), some people can't tell the difference between all these colors. Imagine an alternate timeline where being color-blind was standard and only a small percentage of people could see the full spectrum of colors. The majority of "color-blind" people (who wouldn't call themselves that in this other timeline, of course) would probably think the full-rainbow-vision people were confused about colors.

We have normalized things one way, but our perceptions limit us. Everything we know is the way our minds have interpreted these waves around us. We don't truly see things the way they are; it's our mind's best approximation at that particular time. And yes, thinking this way does keep me up at night.

Feel the Burn: Ultraviolet Waves

While visible light bounces right off us and into our friends' eyes so they can see us, ultraviolet light goes a little farther. This kind of light is able to get under our skin, literally. It penetrates into our skin layers because it has enough energy to do so. And that's the problem. It has so much energy that it can disrupt our poor skin cells while it's there. In some cases it can damage DNA and lead to runaway cell growth otherwise known as skin cancer.

These ultraviolet (or UV) rays are why you put on sunscreen. Just like microwaves, these waves get talked about a lot. We worry about "UV exposure" and look for sunblock that covers "broad spectrum" or more specifically "UV-A and UV-B" rays. UV-A and UV-B are more specific terms for the ranges of waves in the UV part of the spectrum. There are also UV-C rays, but we don't have to worry about them because Earth's atmosphere almost entirely blocks them. In fact, the atmosphere is kind enough to shield us from almost all the UV-B rays, too, but the few that get through are still worrisome, which is why you should slather on that sunscreen.

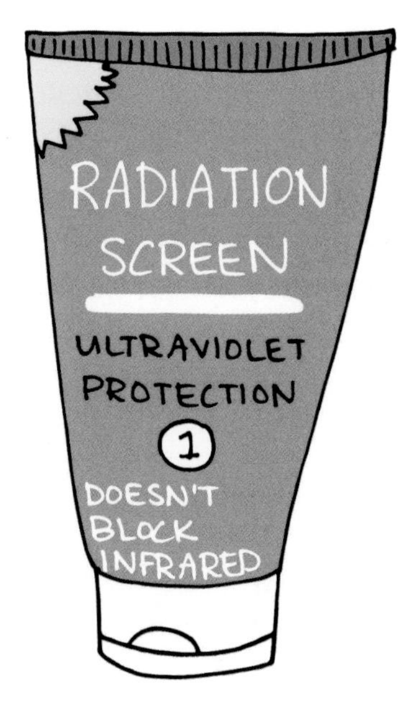

I have a bit of a bias against UV rays. Many of the women in my family have had brushes with skin cancer, so I'm somewhat irrationally angry at these waves. But they don't mean it. They're not sentient at all, of course, but even if they were, it's just their excitement that causes these annoyances to us, like an overhyper dog who piddles on your foot, or a toddler who doesn't know her strength and flails her arms around and bashes you in the nose. (Have these examples happened to me in real life? Yes.)

We can't see these ultraviolet rays. That's part of how they got their name. They're not the violet rays (as in the color purple) we can see; they're a bit more energetic than violet, so they're *ultra*violet. They're the light waves barely outside our visible realm. There are other animals that can see them, though. Some butterflies, bees, birds, and fish can sense this kind of light, where it helps them find food or mates.

I wish I could use ultraviolet light for some purpose. It doesn't seem fair that these animals have access to a whole other form of energy that I'm blind to. But then, since when is nature fair?

Seeing Inside: X-Rays

The last bit of light we'll talk about are X-rays, named after the most mysterious of letters because their discoverer didn't know what the heck they were at first. These waves are most helpful for people writing alphabet books. When you get to letter *X*, it's either those or the xylophone, really. But besides toddler books, X-rays are interesting in their own right.

X-rays are the most energetic waves we encounter with any regularity. There are also, I will mention, gamma rays, but hoo boy those are super dangerous. There is very little everyday about those. Gamma rays are reserved for energy deep within Earth's core and full-on nuclear radiation. There are gamma rays out in space, but they don't reach us here on Earth's surface because they get absorbed by our atmosphere. (We'll talk about that in Chapter 3.) This is great news because gamma rays can kill you.

X-rays are pretty intense, too, but in low doses we can handle them. Dentists use them to look at your teeth, but only every few years, and they won't do it if you're pregnant. I have had to answer that question since I was a teenager. The dental technician escorts me to the X-ray room and says, "Are you pregnant?" "No." "Is there even a slight chance you're pregnant?" "No." "Are you totally sure about that?" "Yes...?"

Interactions like these are a reminder that X-rays, while useful to see teeth and bones, are still to be used with utmost caution. X-rays aren't as dangerous for my adult body cells, but for a developing fetus, a few X-rays could cause things to go haywire.

That's why they put a lead bib over you when you get X-rays, even though they're pointing the machine at your mouth. The lead protects you—or, more specifically, your ovaries or testes—from the X-rays, preventing damage to your reproductive cells. In theory, an X-ray could damage the DNA of one of my eggs, and if that egg got fertilized in the future, it could have an unfortunate mutation leading to some health problems.

The handy thing about X-rays that makes them great for medical imagery is that they go through our soft flesh, but bounce off our hard bones. That's why they produce such beautiful, haunting images. It's like taking a photo with X-rays instead of visible light. Similarly, the X-ray machine at the airport can look right through your bag to see some of its contents. It can't see many details of the softer things in your bag, like clothes, but hard objects like metals show up bright and clear.

There are X-rays out in space, and we would be bombarded with them all day long if our lovely atmosphere wasn't so kind as to filter them out for us. I'm a big fan of our atmosphere's tendency to do that, but some astronomers aren't as keen because X-rays coming toward us are handy for studying faraway galactic blobs and the beginnings of the universe. To do any sort of X-ray detection, we have to send satellites out into space so they can get measurements without our atmosphere getting in the way. Seems like a good compromise to me.

Not Light, but Still Wavy: Sound

After all the waves we've been making in this chapter, you might be wondering about sound waves. But sound waves aren't like the other waves we've been talking about here. They're not light waves, traveling through space at the speed of light. They're vibrations traveling through matter much more slowly. In fact, sound waves *require* something to move through, where light does not. This means that, in every movie set in space where something explodes, the nice loud boom noise you hear is inaccurate. Those sound waves wouldn't travel through the vacuum of space. I think silent explosions are way more dramatic (and a bit eerie), so I don't know why Hollywood types can't get on board. That lot simply can't get over their love of the kabooms.

Sound waves also move much more slowly than light does, which is why there is such a lag between seeing a firework light up the sky and hearing it explode. But similar to the electromagnetic spectrum, there are different types of sound waves, and we are only equipped to hear some of them with our ear holes.

The sounds too low for us to hear are called infrasounds. These are low rumbles, which some animals like elephants can use to communicate between herds many miles apart. Infrasounds are like earthquakes. Strong enough, you could feel them, certainly, but you wouldn't *hear* them.

And similar to the *ultraviolet* light waves we've talked about (which we can't see), we named sounds too high for us to hear *ultrasounds*. And while we can't hear ultrasounds, we can use special equipment to turn them into images, which is useful if you need to peek at internal organs or developing fetuses. You may have heard (ha) that our best friend the domesticated dog can hear some high dog-whistle pitches that we can't detect. And as we age, our ears become restricted slightly at the upper frequencies, so there are some tones that only young people can hear. I can just baaaarely hear some of these higher pitches. Soon they will be gone to me.

There are sounds all around
I cannot hear.

And best of all, we can produce our own sound waves. When I speak with someone, I'm making vibrations that travel through the air to my friend, whose ears then receive and make sense of them. I am very passive when it comes to visible light, simply happening to be there when it bounces off me and enters my friend's eyes, allowing her to see me. But with sound, I'm an active participant, using my larynx to make sound waves.

conversational waves

So the next time you're having a chat with someone, think about how you're generating tiny vibrations in your throat, which travel through the air in rhythmic disturbances like a stone dropped in a pond, so that tiny hairs in your friend's ear twitch when the sound wave hits them, which sends electrical signals to their brain that after years of training and calibration can parse sounds, words, sentences, and actual meaning from a bunch of clicks and tones. But try not to think about it too much or you'll fail to listen to your friend, who will then think you're a distracted jerk.

Such is life.

CHAPTER 3

This Lovely Space Rock: Earth's Atmosphere, Magnetic Field, and Surface

Even when you're sitting quite still, you're flying through space at an astounding rate. You're on a rock—a relatively small one at that—turning as it circles a star. Right now, at this exact moment, all of us here on Earth are spinning around at 1,000 miles per hour (though people at the equator move faster than people near the poles) while we plow through space going 67,000 miles per hour. I'm just glad we manage to hold on.

Like many things in nature, our planet is a paradox. It's so big we can trick ourselves into thinking it has endless resources for us, but it's also so very tiny. Have you ever seen the picture the NASA Voyager I spacecraft took of Earth when it was well past Neptune on its way out of our solar system? The photo is widely known as the Pale Blue

Dot, if you'd like to look it up. That's all we are. A dot. Our splendid space rock is but a tiny oasis in the vastness of space.

I, too, often take our little planet for granted. It's an unfortunate universal habit: the better something is, the more it nourishes you, sustains you, and is always there for you, the more likely you are to completely forget its inherent worth. Earth, your mom, and maybe representative democracy: we take all of these things for granted, but we should instead be grateful for them on a daily, if not hourly, basis. Without them you wouldn't be here enjoying the luxury of reading this very book.

The truth is that many different factors conspired for Earth to become the place it is today. Really, what luck our planet is such a good home for us all. That is, it's good luck in retrospect; if there hadn't been such good fortune, we wouldn't be here to comment on it, but we are, so let's do it.

Take a Deep Breath: All about Air

Take a big, deep, luscious breath. Do you feel that? The air rushing into your lungs? Being aware of your breath is pleasant (unlike being aware of your tongue, which feels like an unsettling distraction) and paying attention to your breathing has a wonderful meditative effect, calming us down. Now, if you want to try out some next-level meditation, think not just about the act of breathing—that handy physics trick where you use your diaphragm muscle to pull your lungs down and suck air in—but also about all the little gas molecules in the air.

The molecule we hear about most is oxygen, since that's an important gas we extract from the air. It exists there as two oxygen atoms bound together. But oxygen makes up only about 8 percent of the air we breathe. The majority of the gas you're breathing into your lungs at this moment is nitrogen. Similar to oxygen, a molecule of nitrogen in our air exists as two atoms bound together. But unlike oxygen atoms, the bond between two nitrogen atoms is incredibly hard to break. Our bodies don't do anything with this gaseous nitrogen, so it goes in and comes right back out. In fact, aside from bacteria, there aren't too many organisms that can do anything at all with these nitrogen molecules.

But anyway, after breathing in some oxygen (and nitrogen), we exhale carbon dioxide. Carbon dioxide is one of the waste products we make. As all our cells go about their daily cellular business, carbon dioxide is a by-product of their metabolism. Just like any clutter, it's best to get it out of the way as soon as possible so it doesn't gum up the works. This right here is an impressive feat of our bodies. In a single breath, the time in between an inhale and an exhale, our lungs are able to do a quick exchange of these gases, grabbing oxygen from the air and doing a switcheroo so that carbon dioxide can be released and jettisoned from our bodies on the wave of our breath.

This is also a reminder that we have something in common with the combustion engines that still run so many of our cars (unless you drive electric, you fancy, eco-conscious, probably well-to-do soul). We both consume fuel, use oxygen, and expel carbon dioxide. (Which on a large scale is a big problem, but we can talk about *that* later in this chapter.)

But it's not as if you grab every last oxygen molecule from the air you breathe. Your exhaled breath still has some oxygen in it. And likewise, you inhale a little bit of carbon dioxide with each breath as well. It's not a complete exchange. This is all fine as long as you still get oxygen with each breath.

Have you ever taken a breath of *mostly* carbon dioxide? It's remarkable how quickly the body picks up on that. When I worked in a microbiology lab, I was often tasked with fetching dry ice, which is frozen carbon dioxide. On one occasion, there was only one slippery slab left in the stockroom, and it kept escaping my clutches. Bending over as far as I could into the waist-high chest freezer, I chased it trying to get a good grip with bulky gloves. I was holding my breath, since the air sitting in this freezer was basically just carbon dioxide gas, but it was taking me so long that I eventually forgot why I was holding my breath to begin with and instinctively drew air into my lungs. I immediately started coughing and choking, my body furious with me. I don't recommend doing this.

So assuming that you're not reading this while free diving or bending over into a chest freezer of dry ice, take in a deep breath of our luscious Earthly air and be happy it has the right amount of oxygen for your cells to continue thriving. They're addicted to the stuff.

Blue Skies and Sunsets: The Physics Behind the Colors above Us

All this air you breathe is part of our planet's truly fantastic atmosphere. I'm not just saying that because I like throwing empty compliments at our planet. Our atmosphere is one of a kind, in our solar system anyway. I'm sure there is a planet out there somewhere in the universe (since it's such a big place) with a similar atmosphere, and we're definitely looking, but either way, our gaseous envelope is simply top-notch when it comes to Earth's goings-on.

First of all, our atmosphere has a nice thickness. This allows it to hold on to a convenient amount of the warmth the sun sends us, and as an added bonus, our atmosphere creates enough friction to make tiny space debris burn up in our atmosphere so we (usually) don't have to worry about falling meteorites and satellite parts raining down on us. That would be so annoying.

A cool side effect of our atmosphere is its apparent color. Take a look outside if you can. (Are you reading this in a bunker?) In between clouds, behold the lovely shade of blue our sky has—from our perspective anyway. You see, this color is truly in the eye of the beholder. It's not like there is something blue floating in the sky. Instead, as light from the sun travels toward us, it hits our atmosphere and gets scattered around by gas molecules like nitrogen and oxygen. Picture throwing a bajillion marbles at an exercise ball (but don't actually do it because, oh my word, the cleanup). Like those marbles would, the light smacks into our atmosphere, deflecting in different directions. During the day, the scattered light that eventually hits our eyes is visible light in the blue part of the spectrum—at least, that's how our eyes interpret it.

Weirder still, people with blue eyes are displaying the same phenomenon. Humans don't make a blue pigment. People with blue eyes just produce very little brown pigment, and when light bounces around their iris, the light that is reflected into your eyes is in the blue part of the spectrum.

Growing up, I didn't understand what "beauty is in the eye of the beholder" meant. It was a confusing sentence to me. I thought it meant that you had pretty eyeballs or something. But now I truly applaud this old axiom. So much of our world is an extension of the way our body experiences it. The sky isn't *really* blue; you pick up on light waves your brain interprets as blue when you look at it. Gaze up right now and think about that.

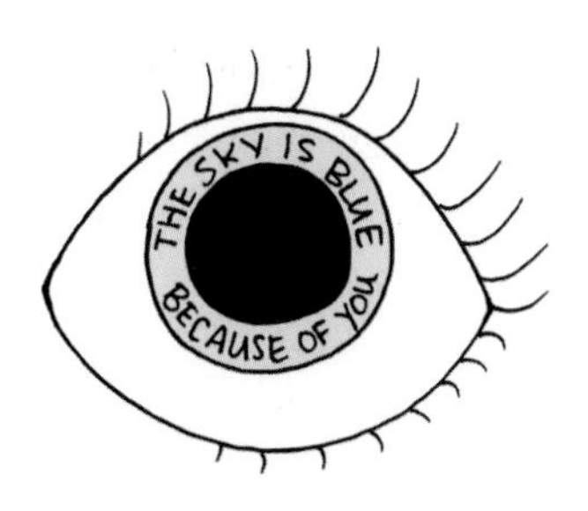

And best of all, the sky turns different colors at sunset. As Earth turns and you're about to move into the shadow of our planet, the light from the sun passes through even more of our atmosphere, scattering extra light. Only longer wavelengths make it to our eyes after all this scattering, which is why sunsets appear as colors in the lower end of the rainbow—red and orange, most notably.

For all the poetry they inspire, sunsets are merely a side effect of the thickness and components of our planet's atmosphere. On a planet with little atmosphere there wouldn't be such colorful sunsets. The sun would simply dip behind the horizon with very little pomp or circumstance. And that would just be boring.

Windy Weather: Why the Air Refuses to Sit Still

If you step outside to watch that glorious light-scattering sunset, odds are you'll feel some air brush against your face. The gas molecules in our atmosphere are constantly on the move, whether it's a light draft or a violent tornado. I don't often think about a nice afternoon breeze beyond "Ah, that feels wonderful," but it's worthwhile to ask, "Why does the air so often refuse to sit still?"

Wind blows for the same reason that you don't plop down right next to the one person on a bus (unless they're your friend). Everything in nature likes to spread out unless there's a damn good reason not to. Air is always trying to even itself out as far as pressure goes, moving from where there is more pressure to where there is less.

But the next logical question is, of course, why are there all these air pressure differences in the first place? It's the sun's fault. When radiation from our star warms our planet, it does so unevenly. Oceans, lakes, mountains, valleys, clouds—these all affect how much an area is warmed by the stream of energy coming at us from the sun.

Gases behave differently when they warm up. The atoms and molecules move around faster, causing them to spread out and rise. That's why the steam from your coffee cup always goes up. And when air is cold, the atoms are slower, which makes a gas contract and sink (which is why the cold vapor coming off dry ice doesn't rise like steam).

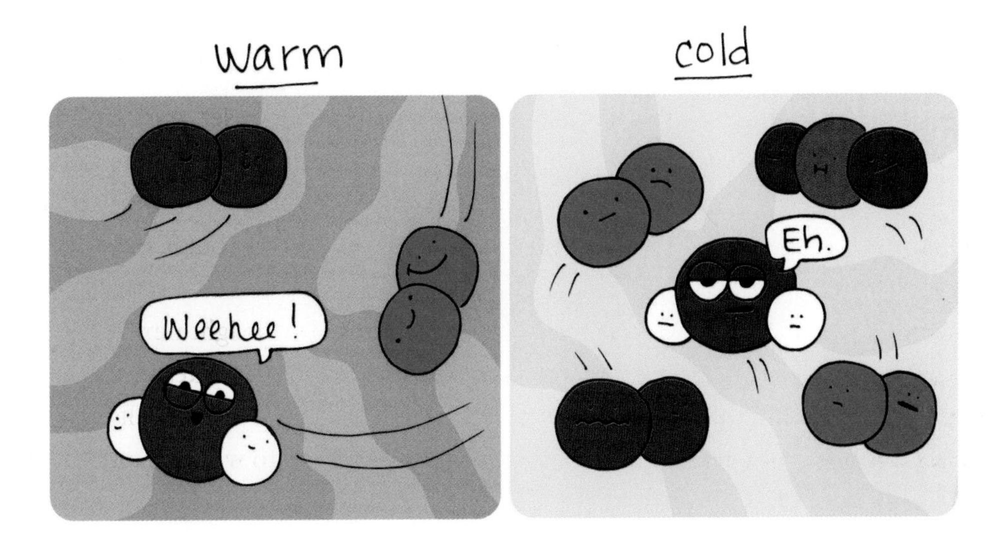

When a pocket of air gets particularly warm and rises, it leaves a low pressure area behind it. And as much as gases dislike being clumped together unnecessarily, they also can't stand a hole (an area with lower pressure and fewer atoms than the surroundings). Ever the problem-solvers, they rush over to remedy the situation.

It sounds like the air is stressed, moving this way and that way under pressure, always trying to even things out but never quite managing it. Things keep changing, and all sorts of obstacles (whether it's a mountain or a tree) impede the air's movement, causing the air to rush around them. These gases are trying to multitask too much.

When you feel the air move past you, it's on its way to fix a pressure difference somewhere over there. Moving past you is circumstantial. If the pressure difference isn't that stark, there isn't as much movement, so we just feel a nice refreshing breeze.

When there are major pressure differences, the air takes it very seriously and lots of air rushes fast to fix the issue. In that case, the wind can cause new problems along its way. For us. This rush of air can knock down trees or, if the angles are just right, start tornadoes.

But even a simple, refreshing breeze is the result of this spinning rock we're on that is warmed by a nearby star. It rustles the leaves of trees and gives something for birds to soar on. It's always on the move, and is quite the overachiever. Good job, air.

Shields Up!: How Our Magnetic Field Protects Us

How did you get to where you are today? Your parents? Your socioeconomic status? Your adorable personality? No. You're here only because Earth has managed to produce a magnetic field around itself. If you want to see evidence of this, you need only hold a $10 gizmo called a compass. It has a little metal needle allowed to freely spin (usually set in water or oil) so that Earth's magnetic field can push it to point toward the North Pole.

I know what a compass is, but every time I think about this, I'm slightly weirded out. All day, every day, there is an invisible force pointing toward the North Pole of our spinning space rock, and it's strong enough to move a piece of metal. If you told a friend that you had a simple device that could receive instructions from an invisible force that surrounds us, they might question the wisdom of maintaining a relationship with you. And while the fact that a compass can always tell you which way is north is stunning all by itself, more mind-boggling are the other ramifications of this mysterious magnetism.

The same force that nudges a compass needle also creates a giant shield around our planet that protects us from outer space. Now, I don't know if you've ever been out in space, but it's not a particularly hospitable place. On top of the fact that there's

no air to breathe and the temperature varies widely, there is also dangerous radiation and something called, rather poetically, the solar wind—a flow of charged particles emanating from our sun that can kill you.

Our lovely planet's magnetic shield deflects this dangerous space stuff, which otherwise would strip away the gases in our atmosphere, leaving our space rock naked and vulnerable, and eventually destroying us all. There are so many gruesome options to choose from when it comes to world-ending catastrophes, but that one would probably be the most unsettling.

The best part is that all of this magnetism, whether it's moving a compass needle or protecting you from space, comes from the inner guts of our planet. The swirling iron-rich rock soup in the outer core of Earth conducts so much electricity as it churns that it turns our planet into one giant electromagnet. This is happening under your feet *right now* (well, 1,800 miles under your feet). Our good fortune in this regard cannot be overstated. It's not a guarantee that a planet stays active like this and continues to protect its inhabitants from the elements. Mars may have once had such a magnetic field, and now it doesn't, which is partly why it's a somewhat rugged place.

The next time you are having a truly craptastic day—one that gets off on the wrong foot because a neighbor's car alarm woke you up at five a.m., and you burned your toast, spilled coffee on yourself, and missed your morning bus—I want you to stop and think about our planet's magnetism. Or better yet, buy yourself a little compass. Even if you don't need to know which way is north at any given moment, it can remind you of the invisible force that is protecting you, and sometimes even guiding you, all day, every day.

Not-So-Solid Ground: The Science Beneath Your Feet

Not only are we spinning on a rock, flying through space, but the ground we're standing on is also moving. Not much, though. Depending on where you are, the seemingly sturdy ground is crawling anywhere from 1 to 4 inches or so per year. Over time, that adds up. If you've lived in the same place for just five years, your home could now be nearly 2 feet from where it was when you moved in.

So why is the land we live on slowly but surely moving? It's because the plates of crust that cover our planet don't stay still. The churning deep inside Earth not only makes a giant electromagnet, it also creates a force that steadily shifts the bits that float on top. There are thirteen main plates on our planet, and they're constantly spreading, smooshing, or sliding around.

We can usually go about our lives not worrying about the ground. We're free to *treat it like dirt* (heh) and *walk all over it* (sorry, I had to), but the occasional rumbling is a reminder of the literally shaky ground we're on.

I live in Southern California, where two of Earth's plates have been sliding past one another for eons. I know I said that the crust moves a certain distance per year, but it's not like it does so slowly, steadily, and predictably. That's an *average*. Sometimes it waits a while and makes a big jump to catch up. When that happens, we call it an earthquake. They're a fact of life if you live near the border of these plates, where the movements deep in the bowels of Earth can cause some crap to happen at the surface.

But the movements of our planet don't just shake the ground from time to time. They also make mountains out of molehills. When these plates press together, some material gets pushed upward, forming vast mountain ranges, arguably the most gorgeous features on our planet. But I'm biased; I grew up in the Sierra Nevada mountains.

Like so many things in nature, the movement of these plates across Earth's surface is a painfully mixed bag. Volcanoes, earthquakes, tsunamis. These are forces of destruction, yes, but they're also forces of creation. These plates make mountains, and volcanoes can make islands like Hawaii. I think we can all agree that's a major bonus.

This constant movement also means that the placement of the continents right now has never happened before in our planet's 4.6-billion-year history and never will happen again. Each moment on Earth is unique and special in its own way, and you get to experience it. Congratulations, by the way.

Fossil Fuels: How Ancient Dead Creatures Power Our Cars *(but Really Shouldn't Because of Climate Change)*

As you go about your business on one of Earth's crustal plates, you probably use fossil fuels on a daily basis, unless you live off the grid and only transport yourself via bicycle. I put gas in my car only yesterday. I stood there, babysitting the pump as it filled my gas tank, staring off into space and thinking about what I was going to make for lunch. I completely forgot to ponder what I was putting into my car at that very moment. The truth is that when you fill your gas tank, you're loading up on the cooked remains of long-dead creatures. And you just use them to drive to the mall, you monster.

Oddly, even though we commonly call them *fossil* fuels, it's easy to forget what that stuff is.

A long time ago, on this very planet, there were ancient oceans full of thriving phytoplankton (algae floating in the water that can photosynthesize like plants). These tiny organisms have been around for millions of years, and they're still around today. But in some places on Earth, between fifty million and 150 million years ago, whole heaps of them died and were buried in mass marine graves. All the molecules that made up their tiny bodies got smooshed together and baked deep underground. And as millions of years went by, they were slowly changed into something new: petroleum.

In a few key places around the world, petroleum is still trapped underground in between different kinds of rocks. For the last one hundred or so years, we've been pretty insistent about finding these reservoirs and pulling the stuff up to the surface. We like to process it and put it in our cars, trucks, airplanes, and power generators. We also use it to make things like plastics. This substance has transformed our world completely in just over a century.

We also know that the waste products fossil fuels make cause climate change and other environmental disasters. These days, most of us are aware we should use as little as possible. I don't go joyriding all day long because I don't want to waste gas and, since some of the electricity in my home comes from burning fossil fuels, I turn lights off when I leave the room.

But even this idea of not wasting energy becomes an unexamined habit. When you unplug the phone charger you're not using, you're saving money on the electric bill, sure, but you're also creating less pollution and conserving the long-dead remains of tiny ancient creatures. This is something we should all be interested in doing because it took over fifty million years for those tiny ancient creatures to turn into petroleum, and once that's gone, we have to wait millions of years to get more of it. And, I don't know about you, but I'm not willing to wait that long to turn on the lights. Perhaps we can find sources of energy that aren't limited to gunk we find in the ground. Wouldn't that be lovely? I bet it would also cause less pollution.

Our Little Space Stalker: The Moon!

Now, I don't know where you are, but there's a fifty-fifty chance if you go outside right this second you'll see the moon lurking somewhere in the sky. When was the last time you stared longingly at its beautiful face? We are very lucky indeed to have this visible floating space rock that follows us everywhere we go.

We're the only planet in our solar system with just one moon. Mercury and Venus don't have moons at all, the poor dears. Mars has two lumpy ones, and the outer planets have arguably too many moons: Jupiter, Saturn, Uranus, and Neptune have dozens (and we keep finding more). Indeed, we have found so many, and some are so small, we may have to discuss how big something has to be in order to be considered a moon at all. Hell, even diminutive dwarf planet Pluto has a handful of moons.

Some moons are captured asteroids, bodies that passed by quite some time ago and got roped in by a planet's gravity. It's like they came to a dinner party and then never left. But our moon isn't one of these hangers-on. Nope, it's the result of a violent space crash many billions of years ago.

As Earth was still forming, something pretty hefty (about the size of Mars) came barreling through space and smashed into baby Earth, sending planet guts out into space. That material—some from Earth and some from the reckless space driver—glommed together and, with the pull of our gravity, has stayed with us ever since, although farther away than you may realize.

How far apart they are.

Yes, it's a friendship that got off to a pretty rocky start, but it's been going well for several billion years now. In some alternate timeline, where nothing crashed into our young planet, we wouldn't have this moon following us around.

And it's lucky it happened this way. As the moon circles us, trapped by Earth's gravity like a ball on a string, it also steadies Earth's wobbly axis, which has a great side effect of stabilizing the climate on Earth, making this planet a more predictable, hospitable place for living things. If it weren't for our moon, we might not be here today.

When you look up at our moon, what do you see? Is it half-illuminated? A thin crescent? Or is it werewolf-provoking full? We have a front row seat to this beautiful monthlong display, but it's easy to forget to look up as the moon passes by, and we can space out about the real reason it puts on its nightly performance.

I'll admit to you that I couldn't explain the phases of the moon until well into adulthood. My education didn't cover any astronomical phenomena past maybe fourth grade, so my ignorance about our moon orbited me until I worked for NASA. No joke.

So let's do this.

The moon, just like Earth (and just like any other body in our solar system), is always half facing the sun, and half in the shade. One side is having daylight; the other is in night. As the moon orbits us, we get various views of its lit face. When the moon is "full," we're looking at the side as it's completely facing the sun. When the moon continues its orbit, we eventually see it half-illuminated by the star at the center of our solar system. And lastly, every month or so we have a "new" moon, where we don't see it much because we're seeing its dark side, as it's passing in between us and the sun.

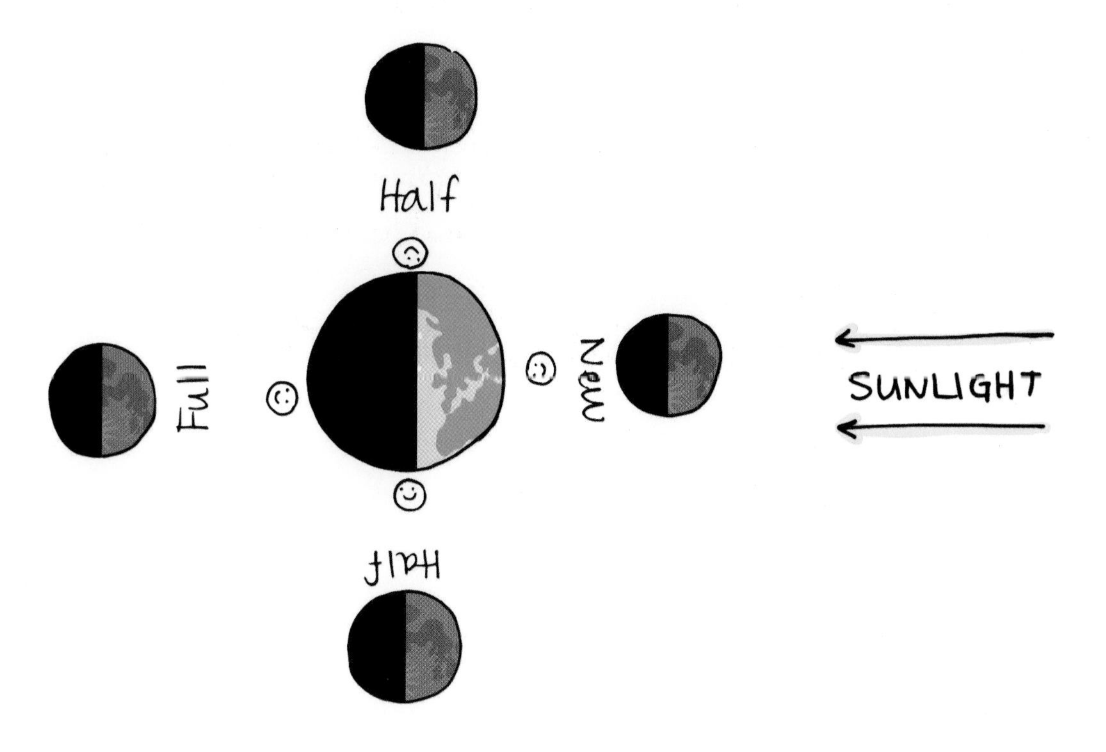

And to quickly clear up a common moon misnomer, you may have noticed I referred to the moon's "dark side." At any given time, half the moon is in darkness, but there is no side that is *always* in the dark. But we do always see the same face of the moon, so while there is no "dark side," there is a "far side." We Earthlings have never seen the moon's backside because the moon is what we call "tidally locked" with us. Even though it goes around us, the moon always faces the same side toward our planet. It's like a creepy stalker circling us and never breaking eye contact. While you're moon gazing, picture what its other side might look like, with a different set of beauty-mark craters.

Now imagine how it would feel to look at Earth, our cloudy blue planet, while standing on the dusty surface of the moon. Against the backdrop of dark space, you would finally be able to appreciate how special and fragile our home is. Unfortunately, moon travel for people like us may never happen, but imagining it is the next best thing.

And while there's no end to interesting factoids and fascinating space physics about our moon (eclipses!), its best quality is that it reminds us of where we sit in our solar system and the universe. Our planetary home, as wonderful as it is, is just a well-established space colony, another vulnerable rock floating in the darkness, half-illuminated by the sun.

CHAPTER 4

The Rocks on This Space Rock: Geology, Earth History, and the Elements We Encounter

We are surrounded by materials that our best friend, Earth, has made for us. Some of them took hundreds of millions of years to fashion, and I just walk right by or set my coffee mug on them without so much as a thought, much less a thank you.

I'm talking about metals, minerals, crystals, and seemingly ordinary rocks. Each of these, regardless of how we have incorporated it into our lives, like the silver in my favorite ring or the paver stone in my front yard, all started with our planet, our marvelous space rock. The churning of Earth's interior, on top of driving the magnetosphere that keeps us safe, also creates opportunities to transform materials.

Though most of the atoms on Earth have been here since the beginning (except for some that arrived here in meteors and comets), over the eons, as material has been heated, squished, heated again, and moved around, it has formed all the combinations that make the landscapes and types of rocks we find in the crust of our planet. To marvel at them, you can step outside and look at the ground, but you can also admire these planetary wonders from the comfort of your own couch by looking at tiles, jewelry, or your many devices, among other things. Let's think about where they all came from.

Rocky Picture Show: The Stories of Sedimentary Rocks

When I think about Earth's crust, the bit that sits right on top of our planet, the stuff of Earth I see every day without fail, I think about rocks. And even though we named a very impressive music genre after them, I still feel as though rocks are a bit unappreciated.

The next time you are able, locate a rock. Even if it's simply a little landscaping stone in someone's front yard. Grasp it in your hand, turn it over, and imagine where it came from. You're holding a little piece of planet Earth right there. And unless it's a meteoroid (which are very rare, and congrats to you for finding one), the atoms in that rock have been on this planet since the very beginning, about 4.6 billion years ago. The very same atoms were in the swirling cloud of dust of our early solar system, the raw material that smashed together and glommed into the planets we know today.

But the atoms in the rock you are holding (or are imagining holding) haven't been in that specific arrangement the entire time. That's the thing about rocks. They're not all the same age. The atoms have been here, but they have been moved around and in different relationships with other atoms during their time on Earth, just like the atoms in our bodies have been up to different things.

Odds are the rock you picked up is something called a sedimentary rock. Most of the rocks on land, which is where I assume you live, are this type. Rocks of this sort were made over vast amounts of time as material was deposited in layer after layer and pressed together, burdened by the layers above it. This is the sort of rock we look for fossils in because as those layers are laid down, sometimes dead animals get trapped in them and preserved.

But what are the layers made of, when you get down to it? Sedimentary rocks started out, as luck would have it, as sediments, small loose particles like sand, silt, and clay. The terms *sand, silt,* and *clay* don't describe the actual atoms that are in those materials, but rather the size of the grains you're dealing with. Thanks to beaches, we're familiar with sand. It's made mostly of quartz, and it has a grainy texture that feels great when you dig your toes into it. Silt and clay are much finer. They could still be made of quartz, just smaller pieces of it.

You would think that we'd be more concerned with the exact makeup of the minerals in sediments when we define them, but it's mostly particle size that matters. And there's a good reason for that: coarse sediments and fine sediments act very differently when they're mixed with water. When water mingles with sand, the pieces are chunky enough that the water doesn't change the texture all that much. Wet sand is still coarse and undoubtedly sandy. But when silt and clay get wet, they turn into mud. The finer the particles, the muddier and muckier you get, sometimes annoyingly so, for me, at least.

I once found myself in the middle of the silty playa of the Black Rock Desert in Nevada during a light rainstorm. Later, the powdery lake bed seemed dry, but as I drove my (rental) car, I found myself in a damp patch of silt. I couldn't tell by looking at it, but I quickly realized I was in trouble. This wasn't simply mud. It was like being stuck in the thickest cake batter imaginable. I spent the night in the car and then hiked to the nearest highway where I regained cell service and called a (very, very good) friend who drove for hours to retrieve me.

But when mud isn't trapping me in the middle of the desert, it can form rocks, given enough time. Indeed, when sediments sit in one area and have layers build up over them, they start to press together, and they can transform into stone. Silt sediments can turn into siltstone. Mud can become mudstone. Sand can become—wait for it—sandstone. I would like to thank whoever made sure these terms crossed over from sediment to sedimentary rock. There's nothing wrong with keeping it simple.

Going back to the rock you found (or are going to find when you can), if it is, as I suspect, a sedimentary rock, think about where its sediments came from. Before those atoms made that rock, they existed somewhere else on our planet, in a different rock, or spewed from a volcano. And somehow, moved by wind, rain, a river, or between the toes of a mole, it got pushed into its current arrangement that you're looking at now.

The rock you have seems so sturdy and permanent, but it's just one of the many lives of those atoms. That little rock someday will get broken down somehow, and it will become small pieces that will go their separate ways, possibly becoming new sedimentary rocks. It will take millions of years for all of that to happen, as old rocks are destroyed and new ones are made, but still, you're holding on to a transient thing on the surface of our planet.

And that's only the story of the one rock you're holding. Think about how many other stones there are out there, each with its own little rocky story, which they're happy to share if we're willing to listen.

Marble Countertops, Granite Floors: The Metamorphic and Igneous Rocks of Fancy Kitchens

Now, sedimentary rocks are great, but let's not stop there. We have other types of rocks to ponder, like marble and granite, the stuff of many a fancy kitchen countertop or museum floor. These two materials are often confused, probably because they're always placed side-by-side in home improvement store showrooms. I used to get them mixed up too. Let's get to know these snazzy rocks better.

Granite is cooled magma, made of lots of different minerals that simmer down and form small pockets of different types. That's what gives it its characteristic colorful, speckled look. It comes in a wide variety of hues depending on what exact ingredients it has, which vary depending on where on Earth it was made.

But whatever color granite is, or how big its flecks, it's cooled Earth guts. It used to be deep underground in a swirling mass of hot rocks. If you vaguely remember your rock types, this one is igneous, which means it's made of cooled down lava or magma. Granite is the most common igneous rock we come into contact with, unless you live in Hawaii and are a mere stone's throw away from some of those volcanic rocks. But since I don't live in Hawaii (as much as I'd like to), granite is the cooled lava I know best.

Marble, on the other hand, is a pressed, layered rock. It has those very pleasing ribbons of color through it, often black streaks in a white-gray base, because in a past life it was a sedimentary rock. Most marble was once limestone, made largely of layers of calcium carbonate from old dead sea creatures whose hard skeletons smooshed together to make a white-ish flaky rock. But marble isn't as crumbly as regular limestone because *this* limestone went through some serious pressure and heat. After being buried deep underground, the layers of limestone, which were already pretty tightly packed, got smothered and compressed. The limestone went from being a cheese sandwich to being a grilled cheese panini.

This type of rock is called metamorphic because it's gone through this transformative experience. But even though it has been through such an ordeal, it still retains much of its identity. It can even hold on to fossils through Earth's panini press. In large slabs of marble, you can sometimes catch glimpses of creatures from long ago.

The next time you find yourself in a fancy bank building or marble-lined museum, take a closer look at the slabs and tiles. Do you see any strange shapes, like cones or squiggles? You might be looking at shells. Those creatures used to be *alive*. They could be hundreds of millions of years old. What a ride they've been on since. After bumbling through an ancient ocean, trying to find food and avoid predators, hoping to meet a mate to continue their species, this creature in one way or another perished and was buried under layer after layer of sediment, to eventually be pushed deeper into the Earth's crust, into a planetary pressure cooker that transformed the rock but left its little skeleton somewhat intact. And then the rock eventually got pushed close enough to the surface for humans to find it and haul it out of the ground. And then it got installed in the building you find yourself in.

Granite and marble are far more than good-looking countertops. These materials give us a peek into the workings of Earth, so in addition to enjoying them for the way they light up a kitchen, take the time to also visualize their journey on this planet and the cooled magma or pressed layers they contain.

Totally Metal: The Source of Our Favorite Shiny Things

Think about the metals of your life, whether it's a piece of gold jewelry, the copper in your smartphone, or the aluminum foil wrapping a slice of leftover pizza—by the way, can I have a bite of that? All those metals came from Earth too. People dug into the crust (the Earth's crust, not your leftover pizza crust) to pull big rocks made of different metals out in chunks, and someone else extracted each metal, melted them down, and molded them into new shapes.

But Earth didn't *make* the metals; they either came with our planetary space dust starter kit or hitched a ride on meteors careening through space and crashing into our planet long ago. So then where did space dust and meteors get *their* metals? It's strange to picture iron, tin, and platinum floating around in space, but that is indeed what happens, especially after a big star explodes.

All the good stuff comes from star guts. The universe would mostly be hydrogen and helium—the two smallest types of atoms—if it weren't for stars and all their farts. Inside those big burning balls of gas, reactions are happening that fuse atoms together, forming bigger and bigger elements. This process doesn't only make metals, as this is also where the carbon and oxygen in our bodies came from (and all

other elements bigger than hydrogen and helium), but so many of the elements on Earth are metals, it seemed like a good time to talk about the star factories where big atoms are made.

Stars have an interesting distribution system that spreads these bigger elements far and wide: they collapse and then explode (the reverse order of what I do when I'm having a rough day). Now, exploding is not the way that *I* would go about disseminating stuff I worked hard to make, but hey, it gets the job done. It's something stars tend to do at the end of their life spans. The sun (our closest star) will die someday, too, sending the elements it forged over billions of years out into space, where they could be incorporated into new solar systems.

Our planet has plenty of metals, but a lot of them sank deep down into the core, never to be seen again, as our planet was coalescing. We can only, literally, scratch the surface of Earth, so we are limited to what we can find up here. And we've decided (or the laws of economics required) that the rarer a metal is, the more valuable it is. But all our notions of value are Earth-centric (like everything else is). Imagine if we jumped through a portal to a planet where the occurrence of metals was the reverse of ours, and they had no shortage of gold but very little aluminum. They might use gold foil to wrap their pizza and give each other aluminum rings to show their deep affections.

Now that we know where metals came from, it might be advantageous to define what exactly a metal is. At least you'd think so. There isn't a cut-and-dried definition for us here. It depends on whom you ask. A chemist, physicist, geologist, and astrophysicist will give you slightly different answers. In general, though, and what you already intuitively suspect, is that a metal is something that is often dense and strong, conducts electricity, and, when polished, appears shiny. There are dozens of kinds of metals, as well as a few *metalloids*, which are things that behave in semimetallic ways. In all, there are more than ninety metallic-y elements, including familiar faces like silver, gold, and lead, but also things like lithium, scandium, and bismuth.

In addition to those pure metals, there are alloys, or mixtures of metals. These don't occur naturally all that often. The ones around you right now, like stainless steel (iron + chromium and a few other metals), cast iron (iron + carbon), or brass (copper + zinc), are the result of humans playing metallic matchmaker. Those atoms wouldn't have otherwise met and assumed this arrangement. But luckily for them, we discovered that mixing metals can make even stronger materials than the individual elements alone. It's a match made in heaven.

As you encounter your daily metals, from your keys to your cookware, think about where those materials came from, how they were made by Earth or by people, and where the atoms came from before they were even part of our planet. It can send you on an atomic space journey without ever leaving home.

Just Passing Through: Why Can We See Through Glass, Anyway?

In addition to your many metals, you probably touch glass on a daily basis but have little time to think about what a weird material it is. Glass is made mostly of the same stuff as beach sand: silicon and oxygen. Strangely, it has the exact same atomic makeup as quartz, but on a very small scale, they're arranged a bit differently.

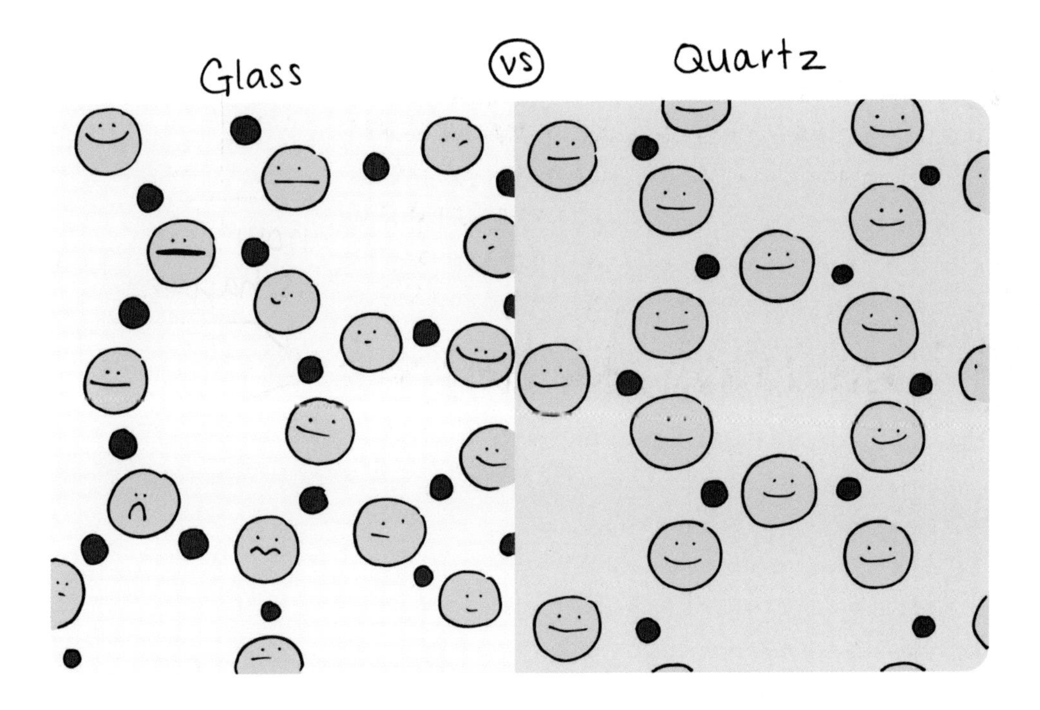

Glass is basically messy quartz. While its ingredients are the same, as it cooled into the shape you are using in your drinking glass or on the screen of your smartphone, it didn't have time to organize itself into the perfect arrangement of quartz.

What's super weird about glass is that it doesn't quite fit the definition of a solid material, but it's not a liquid either. It's what scientists call an amorphous solid or rigid liquid. It simply doesn't have the organized structure of most other solids, and over time the atoms can slightly adjust and settle into a more orderly state. That doesn't mean it flows to the extent that we could see it with our eyes, though. If you've ever heard that old glass windows are thicker at the bottom because over time the glass has moved downward with the force of gravity, that's not true. Glass holds its shape.

But glass's messy structure, or not-so-organized structure, is why it's also easily broken, even in some cases by sound.

The opera-singer-breaking-the-glass trope of cartoonish lore usually has a large Viking-helmeted woman shattering a goblet by singing a high-pitched note. But it's not the upper echelons of the human vocal range that can shatter glass. It's the individual glass object's resonant frequency, which you can find if you rub a wet finger around the rim and make the glass hum. (It's one of my favorite things to do at fancy restaurants and weddings, no matter how many people I annoy.) That pitch, if played loudly enough at the glass for a while, is what can make the glass start to vibrate, wobble, and eventually shatter. If you're patient and can sing the same note loudly for extended periods of time, you could do it, too, no Viking helmet required.

Glass is also special because it's one of just a few materials we use that lets light pass through it without interacting with it. We know this about glass, but we take this very rare quality for granted every time we look out a window. Why does glass allow this?

Back when we were talking about visible light waves in Chapter 2 ("See for Yourself: The Visible Spectrum"), I mentioned that when we see something, we're sensing light waves that bounce off an object. But that's not the only option light has when it encounters something. It can be absorbed by the object, or it can go right on through it. How does the light wave decide what to do? It's up to the material's electrons.

Electrons buzz around the protons and neutrons at the center of an atom, but they aren't all the same. Electrons exist in distinct energy levels, like they inhabit different floors of a building. Incoming light waves have their own level of energy, too, and they only interact with electrons that are on the same floor. Otherwise, they never run into each other.

The electrons of the atoms in glass (silicon and oxygen) live on energy levels that visible light doesn't overlap with. So this type of wave can pass right through glass without interacting with it. However, ultraviolet light—the kind that can burn us—*is* aligned with glass's electrons. Much of it gets absorbed by glass, which is why you don't often get sunburned while sitting inside.

So pay closer attention next time you take a sip of water from a glass. That's a fascinating material you're holding. But don't serenade it, because you might destroy it.

Ooh, Shiny: Diamonds and Other Crystals

One of the most celebrated and coveted of all Earth's offerings is diamonds. In many cultures we give them to people if we think they're special. People have literally died over this crystal. What's the big deal?

The first thing diamonds have going for them is that they're rare. If you go to the nearest patch of dirt and start digging, you're not going to come across a diamond hanging out there. They're hard to find, and they're not evenly distributed around Earth's surface. Diamonds form under very specific conditions deep inside the planet, and the areas where these crystals were made rarely come close enough to the surface for us to get them.

Really though, when you get right down to it, a diamond is just a bunch of carbon atoms. And seeing as how our own bodies are made mostly of carbon, you'd think it wouldn't be such a big deal to have a hunk of carbon, but this element does some pretty interesting things when left to its own devices.

Carbon can form two different types of crystals: diamond is one, and graphite is the other. In graphite (what pencil "lead" is actually made of), the carbon atoms are arranged in rings layered on top of one another. What makes this substance so handy for scrawling shopping lists and sketches of smiling atoms (although that part might just be me) is that these sheets of carbon come off easily onto paper.

But when carbon atoms are pressed together at incredibly high temperatures and pressures—brutal conditions that exist only deep inside our planet—they arrange themselves differently into a lattice that we call diamond, the hardest natural material on our planet.

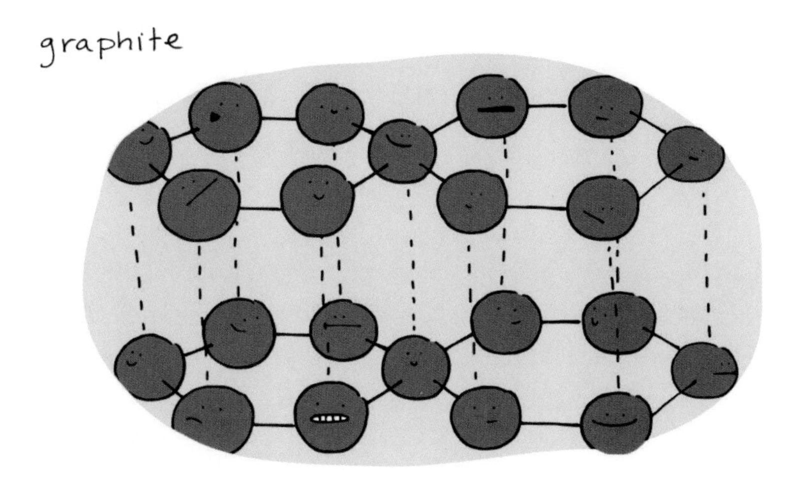

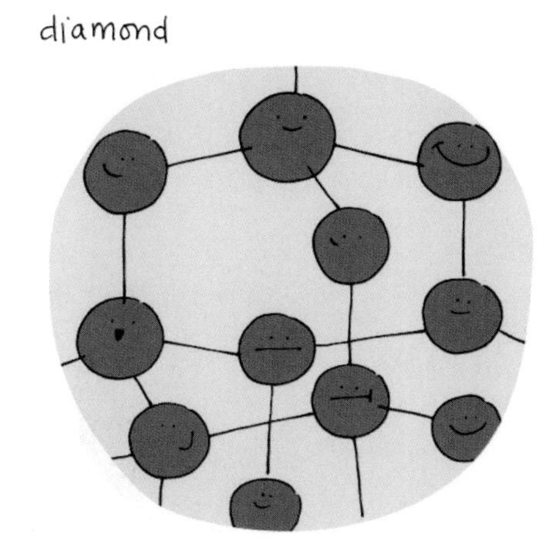

Diamonds take a while to make. We don't know exactly how long, but it likely takes a few million years. The same goes for many other types of crystals. This means that when Earth was young, it didn't have any diamonds, or other fancy crystals like rubies or sapphires. It had the ingredients for them, but it hadn't cooked them up yet. The many types of crystals we can find today are the result of billions of years of our planet mixing together different raw ingredients like a pastry chef uses flour, sugar, eggs, and butter to make nearly endless varieties of baked goodies.

Just think, some crystals are being made deep under your feet right this moment. Most of them we will never, ever see, but sometimes, thanks to Earth's constant interior churning and crustal movements, some of them make their way up to the surface where (with some digging) we can find them.

The Stuff of Cities: Cement and Asphalt

I don't see a whole lot of natural rocks on an average day—not river rocks, marble, granite, or crystals. No, what I see mostly is concrete and asphalt—the stuff of cities. But these, like all the materials around us, had to come from somewhere. So what exactly are they?

Cement is basically a human-made sedimentary rock. The dry cement you can buy is a ground-up mixture of limestone as well as clay and other minerals. When this is mixed with water, poured, and allowed to dry, it forms a very sturdy type of artificial rock that we people love because we can mold it into any shape we want. As you gaze at the sidewalk, think about how its basic building blocks are some of the most common materials in Earth's crust, and consider how they never would have gotten into this arrangement without our assistance.

And think too of the limestone within the concrete. The calcium carbonate that supports the structure of the sidewalk beneath your feet came from things like corals, snail shells, and the exoskeletons of tiny marine algae and other creatures that lived millions of years ago. Their transformation into limestone was one thing, but we have taken that new material and made something else with it. Two rounds of living things (first the snails, and then us) have changed the arrangements of these basic atoms to turn them into something more and more astounding.

And right next to the gray cement sidewalk is a different building material: an asphalt road, also known as blacktop, pavement, tarmac, or rolled asphalt. And just like the limestone in concrete, it can trace its origin back to some dead stuff.

Back in Chapter 3 when we talked about fossil fuels, I left something out. Some of the buried, baked algae that might have changed into petroleum instead turned into what we call asphalt, a mixture of hydrocarbons and some smelly organic compounds and acids. Where you find petroleum, you also find some asphalt. Humans have been using sticky, gooey asphalt to waterproof everything from baskets to boats for thousands of years, and now we use it mostly for roads and roofs.

We've done a lot with these materials, transforming landscapes with them and building roads that connect people to one another. But how long will the cement beneath your feet or the asphalt under your car's tires last? If aliens visit Earth one million years from now, and all *Homo sapiens* are long extinct, what will they find from us? They could unearth evidence of our vast concrete highways and wonder what it was we were doing with them.

Some scientists suggest that we humans have had such a profound effect on Earth that we have ushered in a new geologic epoch, calling it the Anthropocene—the age of humans. This is no small thing. We divide up the history of Earth based on world-changing events such as mass extinctions and dramatic climate changes, like the one that ended the dinosaurs (with the exception of the ancestors of birds) at what we decided was the finale of the Cretaceous period. A new geologic interval is usually defined by a catastrophe, so it's not exactly a good thing that we might define the beginning of a new one.

Millions of years from now, concrete slabs could be used to demarcate this new regime, the one in which humans, not Earth, made rocks. I hope that's the only signature we leave on the planet, and whatever intelligent species of living thing that exists in millions of years doesn't think we were too weird because of it.

CHAPTER 5

Just a Blade of Grass: Photosynthesis Is Pretty Bonkers

Did you know that you're here only because plants are so talented at using sunlight? Every bite of food and breath of air you take nourishes you because of plants. If all these leafy things threw in the towel and gave up on this whole enterprise, we wouldn't be able to make it without them.

These quiet, green beings are deceiving to say the least. So many of them seem to be passive at first glance. We even say that the ultimate in boredom is watching grass grow. But have you ever stopped to look at a blade of grass before you step over it, sit on it, or whatever it is you're up to today? It's doing something that I would think was impossible: using light to make its own food. The latest technology and complex manufacturing facilities make solar panels that don't do nearly as good a job at capturing the sun's energy, while each little blade of grass (and the millions of other species of plants) do it all day without even stopping to brag about it. We should all be so humble. Here you'll learn where plants get their energy, why weeds are so impressive, and how plants defy gravity.

How Do Plants Do What They Do?: What Photosynthesis Really Is

Unless you're in the middle of Antarctica (and if so, congratulations on making it to the South Pole), there are probably some plants around you. Trees, shrubs, grasses, cactuses—I don't know where you live. But all plants are remarkable.

At a very early age we learn that plants seemingly need just a few basic things to get on with this whole "being alive" business: a place to grow (the ground is often sufficient), water, and sunlight.

But what are plants doing with that radiation from our nearest star? Something we don't often take the time to deeply appreciate. The very same light that you and I can see—that is, visible light—provides energy that plants use to build their bodies. This will take some time to unpack. We call what plants are doing "photosynthesis," mostly because no one thought of a better name (oooh, foreshadowing). While many of us are familiar with this word, it's one of the most misunderstood science concepts. If we split it up, it sounds simple enough: *photo* means "light." Cool. And *synthesis* means "making something."

The plants around you are doing this thing right now. They're photosynthesizing.

Now, let's examine what those plants are up to out there, starting at the very beginning of this process.

At the center of our lovely solar system, the sun is sending out heaps of radiation. That energy goes in all directions into space, hitting other planets, moons, and asteroids, while some of it steers clear of everything and leaves our solar system entirely (allowing aliens on faraway planets to see our sun as a twinkly star in their night sky). But some of it heads straight toward us here on Earth. It goes 93 million miles (give or take, since at different times of the year we're a bit closer or farther away), and some of it bounces right off our atmosphere, deflecting out into the void of endless space. But some of it gets through, shining down on Earth's surface, where a few plants and I are enjoying a nice afternoon.

This radiation is so intense that I can't look directly at the sun (nor should you), and a mere thirty or so minutes of exposure without sunscreen will burn me. But plants stay outside *all day long*. Sometimes on a particularly hot, sunny day I look out my window, and while enjoying the air conditioning I look at a plant and think, "You're hard-core."

But that's not the best part. Not by a long shot. Some of that light from the sun happens upon a little blade of grass (or the leaf of a rose bush, you know, any old plant) and something truly awesome happens. It jump-starts a reaction within a plant cell that allows it to *build* things.

Keep in mind that in nature, making things is hard. I mean, it's difficult in any context (have you ever tried to build even the simplest spice rack?). It's way easier, and often more fun, to destroy things—that is, break them down. We humans, like all animals, excel at wrecking things. And I'm actually not talking about the havoc we wreak on the planet; I'm just saying that we need to find a ready source of energy around us so that we can break it down and harvest it. That's why we have to eat. We are no better than vampires or zombies: we can't exist without killing other living things and sucking the life out of them. (We only abhor these monsters because they do it to *us*.)

But plants are totally different. They don't eat anything. (I mean, yes, sure, there are carnivorous plants like Venus flytraps, but most plants don't do that.) Plants *make* their own food. Not the way that I "make" myself a sandwich. No, plants build their own food supply from the rawest of materials available to them, carbon dioxide and water, using sunlight to run the food-production engine. Think of plants as tiny solar-powered factories.

What it all boils down to is that plants figured out a way to use sunlight to *split water*—no small feat, by the way. I mean, have you ever done that? No, water likes being itself and isn't going to change just because you ask it to. But plants manage to pull off this chemical feat without any expensive lab equipment. They absorb electromagnetic waves and redirect their energy to split H_2O apart, liberating the hydrogen and oxygen atoms in this molecule and releasing energy. That's the key.

The energy from breaking water apart is what ultimately runs the botanical factory. It keeps plants going and, in turn, has allowed us to survive here, too, partly because a side effect of this water splitting is the release of oxygen gas, the very same oxygen you and I breathe, a lucky stroke we'll talk more about later in this chapter.

Now, I know it's weird to think about water holding on to a lot of energy. We think of water as an easygoing substance. But in each individual water molecule, hydrogens are held to oxygen with chemical bonds, and they store energy. When you break those bonds, it's like a tiny bomb going off; the process can be compared to the way that our cars run on the energy stored in petroleum's bonds.

With the energy released from splitting water, plants have the strength to stitch together carbon atoms, making a ring we called glucose, a very versatile building block. Plants can stack glucose together to make cellulose (which in our diet we call fiber), starches (think potatoes), sugars (things like apples), or they can break the glucose right back down again and get energy from it. That's what I mean when I say they "make their own food." They used sunlight to make glucose, which they can then "eat."

In the end, photosynthesis means that energy from outer space is put through a little cellular machine and comes out the other end as energy stored in chemical bonds that living things can use. And to think you were going to sit on a patch of grass without reflecting on this.

Plants Make Do with So Little: Water, Sunlight, and a Breath of Fresh Air

Picture everything you'd need to pack to go camping for two nights. We modern humans require so many things. Tents, sleeping bags, firewood, food...ugh, I'm bored already. Let's just stay home. It's so much better to be a plant; they pack light and can get by with so little that it makes me jealous.

In the previous section I mentioned what plants *do* require—a place to grow, water, and sunlight. But the unspoken requirement is air. Plants can't grow without it. More specifically, they can't grow without the carbon dioxide from that air. This is perhaps the most counterintuitive part of this whole photosynthetic picture. Plants get carbon—the main ingredient of their bodies—from the surrounding *air*. It sounds ludicrous when you put it that way. We often think of air as being pretty much a big nothing. If you call someone an airhead, it's not usually because you are saying they have lots of carbon dioxide plants can use.

While air is definitely less dense than a rock or a bucket of water, it still has mass. Anyone who has experienced a tornado or leaned into a strong gust of wind can tell you that air most definitely has some heft to it. When it moves quickly, it can tear roofs off houses, for crying out loud. And we can see that carbon dioxide has mass when we pick up a hunk of dry ice, which is solid CO_2.

Now, think about the biggest tree you've ever seen. (Was it *the* biggest tree—2,000-year-old General Sherman at Sequoia National Park? I've seen that one. Not to brag or anything.) That tree built itself. All by itself. It didn't suck the life out of

other creatures to do it. It manufactured its trunk, its bark, its branches, its leaves. It made it all out of carbon dioxide and water. If that's not incredible, I don't know what is.

Photosynthesis might be basic biology's best-kept secret. Any time we set out to learn about it, we get lost in the weeds (ahem) of this process and forget what it really is. I was no different. I went over all the nitty-gritty details in school, memorizing the intermediate steps of this complicated process, but I often lost sight of the bigger picture.

The real problem is that the word *photosynthesis* doesn't even begin to capture how bonkers this process is, and what an innovation it remains billions of years after it was perfected by Earth creatures. Photosynthesis needs to embark on a serious rebranding exercise. We've let this go on for too long. Photosynthesis should from now on be called Sunlight-Water-Bomb-Sugar-Making. But I don't know; I'm not so good at this marketing stuff. Regardless of what we call it, this is an ingenious system, exploiting very convenient resources. We're not in danger of running out of carbon dioxide anytime soon (in fact, we have too much of the darn stuff), and sunlight is readily available (you know, until our sun dies in several billion years).

However, plants don't use every bit of the light coming from the sun. When they're photosynthesizing (er, um, Sunlight-Water-Bomb-Sugar-Making), plants exploit particular parts of visible light. They're quite fond of violet, blue, and red light, as those are the main colors that plant pigments like chlorophyll absorb. But what about green light? Well, plants don't use those wavelengths. Chlorophyll reflects them.

Strangely, plants appear green to us only because they discard green light, and then it hits us in the face and registers in our brain. It's what they use least, so it's slightly odd that that's how we know them. But hey, you can learn a lot about someone based on what they throw away, like getting to know someone by looking through their trash. Not that I'd recommend that method.

Oh, and Plants Defy Gravity: How Water Travels from Root to Leaf

Since a key ingredient of Sunlight-Water-Bomb-Sugar-Making, or photosynthesis, is water, have you ever wondered how all that liquid moves into a plant? The roots are pulling water from the ground, but how does the water get up there to those leaves where all the photosynthesis is happening? In a blade of grass this feat isn't as impressive, but think about a tall tree. How in the world does such a plant manage to get enough water high up to the leaves? I mean, it's not like the tree has a water pump.

This phenomenon rivals photosynthesis itself for sheer awesomeness. The water defies gravity and moves up through the tree because of some of water's inherent characteristics.

Two main qualities of water enable it to move through plants this way. First, water evaporates, and, second, water "likes" itself, which we'll talk more about in the next chapter. But for now, I'll mention that water is sort of "sticky" in the sense that there is a weak force that attracts water molecules to each other. Within the trunk of a tree, or the tissues of any ol' plant, there are water molecules all lined up, holding "hands" in a sense—where their hands are this weak force between them called hydrogen bonds. As water evaporates from the leaves, the molecule that's leaving slightly tugs on the one behind it in this big ladder, so the effect is a water molecule escalator, slowly moving liquid into the plant, reaching all the way up to the leaves so the plant can keep doing its thing.

And this all happens as we walk by, totally ignoring that little blade of grass or stalwart tree. We're so self-centered, and we rarely thank these plants for perfecting this process that gives us food to eat and air to breathe. The truth is that animals like us have been riding the coattails of plants for millions and millions of years. It's pretty unfair. But I have no intention of changing this system. It's been working pretty well, you know, for *me*.

Growing Like Weeds: Why Weeds Are So Low Maintenance (and Sometimes Annoying)

Because I applaud plants so very much, a few years ago my husband and I decided to become backyard farmers and grow some herbs and vegetables. Cilantro, potatoes, tomatoes, lettuce, and corn. It went, um, okay. Wanting to be horticultural overachievers, we grew many of the plants from seed, tending to them dutifully and trying to provide optimal soil and water.

Outside, only some of them fared well in the Southern California sun, even with daily watering. After working particularly hard to simply keep these plants alive (as I no longer cared if they produced a single bite of salad), I was walking back inside one day when I noticed a large weed growing behind a piece of patio furniture. With nothing more than a small crack in the cement, this plant was positively thriving. I was enraged. Here I was bending over backward (well, really just forward) to do everything in my power to nurture my underachieving plants, and this weed was like a slap in the face, flaunting its success in front of my vulnerable tomatoes. What an inconsiderate plant jerk.

I shouldn't take things so personally, though, because everything is relative in terms of what makes a plant a "weed." Weeds are basically plants that don't need us, defined mostly as those we didn't put in the ground ourselves in a certain spot on land we consider "ours." But if plants could talk to us about their self-awareness, surely none of them would consider themselves a weed (just like few humans think they're bad people). This categorization is a strictly human concept. Weeds would probably be baffled by our frustration with them, wondering why we insist on plucking them from concrete seams and the edges of flower beds. From their perspective, we are monsters.

Depending on where you live, some "weeds" are just hearty local plants, and they show up everywhere because they're so well suited to the landscape. We resent them simply for their success. A gardener I know prefers the term "plants out of place," but I didn't have the heart to point out that the acronym for that is POOP. Some weeds, however, are *invasive*—meaning they're plants that were brought to an area by people and are running amok.

We humans have a nasty habit of moving living things to new places (and then we call them introduced species). We usually know better than to do this with an-

imals these days, but many people do it all the time with plants without thinking twice. Look around your neighborhood—which of those plants have been there for thousands of years, and which ones did people plant recently because they look nice? In most nurseries, you'll find species originally from other continents that were grown on farms hundreds or thousands of miles away, often in completely different climates, so they might not even be suitable to survive in your area. Or, worse, they might be poised to flourish and become an invasive nightmare, spreading to new areas, crowding out other plants, and competing for their resources. It's like *Botanic Park*.

But whether a weed is a robust local plant or an invasive newcomer, it's impressive that it knows exactly how to survive, and doesn't need a single thing from me. While it's frustrating at times, I still admire its spunk. So, the next time you pull a weed, think about things from its vantage point. I did this the other day. My yard has a weed with spiny barbs all along its stems and leaves, and as I was ripping it from the soil, I cursed it for stabbing me through my gloves. But then I thought about how I was actually committing plant murder, so I figured we were even.

The Food You Eat: How Every Bite Ultimately Comes from the Sun

Take a bite of an apple, or munch on the greens in a salad. Or just have a piece of toast. Think about how every bite can be traced back to the sun. Even if you ate nothing but animals (which is not a great plan, by the way), you'd still be relying on photosynthesis because those animals needed to eat plants as they grew. And don't think you can escape this rule by eating fungi like mushrooms. They live off the remains of dead things in the soil that in life used the sun too. But among those plants that you do (and should) eat, do you know what part of the plant you're eating?

Some of them are recognizable. The greens of a salad are obviously leaves. There are plenty of leafy things we could eat, but we usually go for the softest ones: spinach, romaine, arugula, you know the drill. They are crispy and thin, and we can chew on them with some success. And when you grab a forkful of salad, you are eating the cells that carried out photosynthesis when the plant was alive. The chlorophyll that absorbed the sunlight, the cellular membranes that helped make sugar. In life, they were doing the most biologically extraordinary thing on planet Earth. Now they're lunch.

We also eat stems. Celery is pretty recognizable as the stem of a plant, as is asparagus. When you cut these stem-y vegetables in half, notice the lines that carried water up the plant when it was alive, defying the force of gravity here on our planet. Broccoli is sort of a stem, but we mostly eat the flowers at the top of that plant. The same goes for artichokes; those are big (and delicious) flowers.

And then there are starchy vegetables like potatoes, radishes, and carrots. These all grow under the ground as part of the plant's root system. Many of these are energy stores the plant makes to get through hard times, not knowing that the hardest of times would be when we rip them from the soil and throw them in a pot of boiling water.

Lastly, and probably the most popular, are the fruits. Did you know that all fruits start out life as flowers? It's true. A flower waits for passing pollen—whether it rides

The Horrifying Transformation of a Strawberry Flower

the wind or the fuzzy underside of a bee—and the pollen reaches down and deposits plant sperm inside the flower where the egg waits. Yes, plants have sex too. The fertilized egg (or multiple eggs) eventually turns into seeds, and the ovaries swell and transform into fruits that surround and protect those seeds.

There is an absurd variety of fruits made by this diverse group of plants. Some fruits have the seeds on the inside, buried under lots of flesh to protect them, such as peaches. Others have the seeds all over, like strawberries. And many of the things we consider vegetables are fruits in disguise: zucchini, squash, and tomatoes. This can be confusing because we tend to think of fruits as any produce with a noted sweetness, but anything you eat that has seeds is technically a fruit. And when you eat one, you're murdering little fruit babies. You're despicable and should turn yourself in to the authorities for your crimes against plantmanity.

Just kidding. It's really okay because plants long ago tacitly agreed to this arrangement. Plants produce fruits partially so that animals like us will eat them and disperse their seeds to new areas. Without fruit eaters, seeds could only go so far as a drop and roll from the tree (or bush or whatever) would allow, and as we all know, the apple doesn't usually get very far.

We humans didn't stop there. We eventually planted some of these seeds in the ground ourselves and tended to them as they grew. In fact, with all the energy we spend caring for these plants, it makes you wonder who is truly in charge. At times, it seems like we're their servants.

And some plants wouldn't exist today if we hadn't taken them under our wing, such as avocados. Those enormous seeds—so big we call them pits—are too big to fit down the throat and digestive tract of any but the biggest animals. The only creatures that probably ate them with any regularity were giant sloths. Those enormous animals went extinct about ten thousand years ago (probably in part because people hunted them to oblivion). Since avocados seem to have specialized in just one animal, without those sloths, avocado trees might have gone extinct, too, if the same people who killed their seed-dispersal species didn't find their fruits so delectable and begin cultivating them.

The next time you have guacamole or smoosh an avocado's green flesh all over a piece of toast (with some salt and hot sauce on top, drool), think about how you're eating a once sloth-dependent, human-reliant Ice Age relic. Delicious.

Photosynthesis in the Sea: The Algae That Give Us Oxygen to Breathe

Okay, so you now know that every bite you eat can be traced back to the sun, but so far we've talked mostly about ecosystems on land—the plants, animals, and fungi there that we eat. But what about a bite of salmon, shrimp, or clams? How can you trace these marine bites back to the sun?

Don't worry; I'll walk you through it. Let's first get acquainted with the plants of the sea—algae. They're not true plants by some definitions, since they don't have roots or meet other arbitrary rules we set for classifying something as a plant. But hey, they still do photosynthesis, and that's what it's all about. You might be familiar with seaweed, which is a type of algae, but there are also untold numbers of microscopic critters floating in the open water, making their own food with the sun's energy. These form the basis of the big food chains in the ocean.

That bite of fish? That fish ate algae as a wee baby fish, when it was too small to eat anything else. And some fish eat algae and other plankton (small stuff floating in the water) no matter how old they get. Other fish, the big sporting ones like tuna, eventually eat other, smaller fish. But no matter their size, they all depend on the algae in the water at the bottom of the food chain. The same goes for shrimp, which live off algae and other bits they can scavenge. And stationary critters such as clams, mussels, and oysters filter things from the water, including, once again, algae.

Almost nothing could live in the ocean without algae establishing the lowest level of these food pyramids. (Although, there are ecosystems on the ocean floor that use hydrothermal vents for energy instead of the sun.) The can of tuna in your cupboard is just as tied to the sun's energy as your spinach salad was. None of it would be here if there weren't loads of tiny cells using the sun to make their bodies. By some estimates, half of the oxygen in Earth's atmosphere is produced by plant-like creatures that live in the sea. Half! Those little guys have a big impact. And over time, they have changed our entire planet.

Billions of years ago...

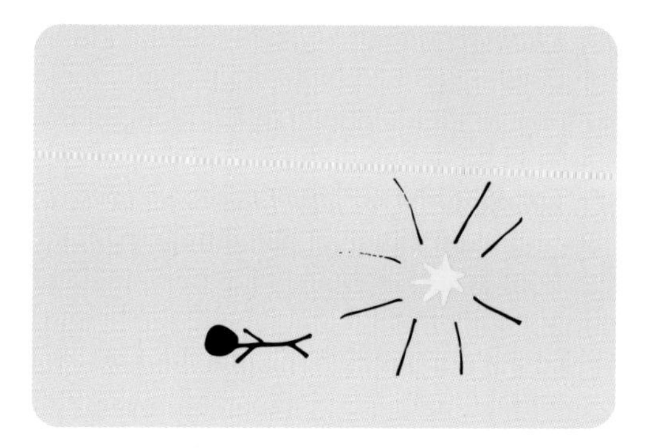

If you hopped into a time machine and went back to see the first living cells around 3.8 billion years ago, you'd have to bring some oxygen tanks with you. There was almost no oxygen in the atmosphere at that time, so you'd immediately start suffocating.

Over billions (yes, with a *b*) of years, untold numbers of these tiny photosynthetic cells spit out so much oxygen that it transformed Earth's atmosphere and made way for hungry animals like trilobites, fish, crocodiles, turtles, dinosaurs, birds, and then mammals like us. In an alternate timeline in which cells didn't figure out how to do photosynthesis, life as we know it never would have happened. Every oxygen molecule you have and will ever inhale is here because of plants. They're the only reason you're not suffocating right this minute, so thanks, plants!

What *We* Do with Light: It's Time to Make Some Vitamin D

I'm sorry to report that we humans definitely don't make our own food with light. But that's okay. We do make something *almost* as cool: vitamin D. And that might sound like a weak consolation prize or the ravings of someone trying to convince you that fat-free ice cream is delicious (gag), but don't be like that. Vitamin D is very important for us, as it helps us grow and maintain strong bones. Without our mineralized skeleton propping us up, we'd be blobby meat sacks. So even though it doesn't nourish us per se, in a way the sun makes us stronger.

But how does it work? Well, we have this molecule we call provitamin D in our bodies. This part we make ourselves, but we need sunlight to help us transform it into the vitamin D molecule that helps us build our bones. Specifically, we need UV-B rays. These otherwise dangerous light rays strike provitamin D in our skin and break one of the bonds in this big molecule, giving our cells the chance to change its shape and turn it into Vitamin D, which can then help us absorb calcium from our diet and keep our bones strong.

This is one of the main reasons it's inadvisable to live in a cave and never emerge during daylight hours; otherwise I'd totally be supportive of that. But really, even staying inside a normal abode with the shades open all day isn't enough because glass windows filter out most UV-B light, like we talked about in Chapter 4. So I'm sorry, but you have to go outside sometimes. It's important for your bones. And social skills.

But don't be out there too much without adequate sun protection either. You might remember that UV-B is the most energetic (and therefore dangerous) radiation that reaches us here on Earth, so while we need a little bit to make vitamin D, too much can cause skin cancer. What's that old adage? "The dose makes the poison." There is an optimal amount of sunlight, as well as coffee, friends, social media, news, chocolate, and butter. Life is about finding and maintaining all of these balances simultaneously. That's why it's notoriously difficult.

So when you do go outside and the sunlight cascades onto your skin, feel the infrared radiation's warmth, and try to picture the UV-B rays diving underneath your top layers of skin to snap a bond between two atoms in a molecule of provitamin D, and how your body uses that opportunity to make a molecule you need to have healthy bones. Go on, stare at your skin while it does this. No one will notice.

And if you absolutely insist on living in a cave, you can probably just buy some vitamin D supplements online or something.

CHAPTER 6

Water, Water Everywhere: The Properties of Water That Make Life Possible

I use the word "need" pretty loosely. I "need" Wi-Fi. I "need" coffee (well, maybe this one does fit the true definition of need). I "need" mascara. But I really do *need* water. We all do. It's one of the exceedingly few things all living things on Earth have in common. Water is so incredibly useful, we can't quite wrap our heads around a possible living thing that doesn't depend on it, which is why the search for life elsewhere in our universe is mainly focused on finding planets (or moons or whatever) with liquid water.

Something I forget, because I can easily get water from my faucet, is how quickly I would die without it. Just a few days is all it would take, which is an eerie thought. I like to think of myself as a fiercely independent woman, but my crippling dependency on this substance is absolutely inescapable. I have to be near a ready source of it, or I have to remember to lug it along with me. Sometimes I'm a bit resentful.

But all that can be forgiven, for on a hot day when thirst kicks in and my brain urges me to drink water (since I apparently wouldn't do it otherwise), there is nothing more pleasant than feeling the cold molecules of water rush over my tongue. What is so special about this substance that all conceivable living things need? Let's find out.

Water Has High Self-Esteem: How This Molecule Is Attracted to Itself, Seriously

Water, unlike most people, likes itself. This is not a guaranteed quality in a molecule. Oftentimes chemicals are perfectly neutral and nonchalant about themselves, neither particularly liking nor disliking themselves. But water is dripping with self-confidence.

The power of water, and its inherent narcissism, lies in its structure—the precise arrangement of its atoms. We touched on this a bit in the context of handwashing (in Chapter 1) and the movement of water up through a plant to reach the leaves (in Chapter 5). Water, while overall neutral, has a positive side and a negative side. The difference in charge comes from the protons and electrons of the three atoms that make up a water molecule. There's an oxygen atom in the middle, flanked at angles by two hydrogen atoms. They stay together in this bond because they are sharing electrons with one another, but the electrons are not shared equally.

The oxygen atom not only has more electrons than the hydrogen atoms to begin with, but it's also more keen to hold on to them. The electron that hydrogen shares with oxygen spends the majority of its time with oxygen. And on top of that, oxygen has two pairs of outward electrons that aren't tied up in a bond. With so many electron-y things happening, oxygen winds up with an overall negative charge. And as a result, the side of the water molecule with the hydrogen atoms winds up having a slight positive charge.

That means that each and every individual water molecule floating in your glass of water, or frozen into place in a block of ice, is like a very small magnet, with the negative end of one water molecule attracted to the positive end of the next water molecule.

The allure between any two water molecules is very, very small. It's not like a strong magnet clinging to your fridge, holding up your grocery shopping list and vacation photos. The force is oh so tiny, but when you multiply that force by the gajillions and gajillions of water molecules in even just a small glass of water, it starts to add up. It's a grassroots effort, this force. The result is that water is a substance that is a little bit more than the sum of its parts and all those small attractive forces make water stronger.

On the scale that we can see, the overall effect of this tiny bit of attraction between water molecules is that water becomes cohesive. It likes being around itself, which also means it forms pleasing little droplets. It's hard to appreciate this because we aren't familiar with too many other liquids, so we don't have much to compare water to. But looking at oil can help.

Whether its olive, canola, or lard, oil doesn't like or dislike itself. It's sort of apathetic in that way.

Think of the last time you put oil on a skillet before scrambling some eggs. The oil spread out nice and easily, right? It didn't put up a fight. It's happy to oblige this request. Now, try and do that with water sometime. Attempt to convince a teaspoon of water to spread out evenly over a frying pan. It cannot be done. The water will always coalesce into clumps, no matter what you do. It respects itself too much to be spread so thin. We could all learn a lot from water, quite frankly.

Solid, Liquid, Gas: How We Experience Water

While water likes itself plenty, we like it even more. Water is so important to us that our entire concept of temperature is built around it. Whether you use Fahrenheit or Celsius, the scale is set by water. When we say it's "freezing" out, or the weather is "boiling"—we're talking about water's freezing point and boiling point (but in the latter case we're being hyperbolic). We simply don't care about silver's freezing point, or oxygen's boiling point in our everyday lives.

Water is the substance that most concerns us. Not only is it a big component of our own bodies as well as our food, but it's in the air around us, and it falls from the sky sometimes. Water also changes phases right before our eyes several times a day. It freezes, melts, vaporizes, and condenses. We experience water in all three of its forms—solid, liquid, and gas—just about every day. (Although, you can go a day without seeing ice here and there, of course, like a hot summer day when you don't open the freezer.)

My favorite of these phase changes is vaporization, when liquid water transforms to water vapor and floats away. I love this process because it dries my towels. Think about it. When you wipe your hands with a bathroom towel after washing them, the fabric doesn't stay wet forever. It slowly but surely dries. We know this, which is why we (usually) put towels on hangers. But how odd that our terrycloth habits are really intended to promote the gradual escape of water molecules that have changed phase over the course of the day.

Imagine for a moment that water didn't do this on its own, and it required effort on our part every time. Dishes wouldn't air dry. Towels would be damp for eternity. You'd spend a lot more time blow-drying your hair. It would be a soggy nightmare. Luckily this is not the case, for water molecules, when left alone even at room temperature, will slowly change into water vapor, and the water that you once used to wash your hands will eventually be in the air around you. You could even breathe it in at that point. I normally forget about the water molecules once they've left my towel, but they're still in my home. It's like they're haunting me.

We can, if we choose, heat water to the point that it literally can't contain itself and leaves its liquid form in droves. This, my friends, is known as boiling water. I achieve this state by burning natural gas in my stove or using electricity for my water kettle to contribute so much heat to a few cups of water that the molecules get so excited they cook my pasta and steep my tea.

When I set out to make pasta when I'm already starving, I get frustrated by the amount of time it takes a pot of water to boil. I stand there staring at the stainless-steel container, willing the water to reach its boiling point. It takes a lot of heat to raise the temperature up to this, but what is the significance of water's boiling point? It's the temperature at which the water molecules can readily overcome the surrounding pressure on them.

One of the reasons water exists as a liquid at all is that a force is pressing down on it. You might look around and think, "What force? There's nothing here." The force is barely noticeable to us, just like the matter exerting the force. It's air pressure. The entire column of air of our atmosphere is sitting on top of that pot of water, helping keep the molecules pressed together enough to make it continue to be liquid water. But adding heat gives the water the energy (and bravery) to overcome the pressure and leave the confines of the pasta pot.

We don't usually have to think about it because the air pressure on Earth is relatively constant. But if you tried to make macaroni and cheese in a vacuum chamber (with no air pressure in it), you'd get frustrated pretty fast. Without air, the water molecules would happily and easily boil without any additional heat because there would be no air pressure weighing them down, forcing them to exist as liquid. This is also why cooking things at high altitudes can take longer. With less air pressing down on it, water boils at a lower temperature.

But there is more to water than vaporization and boiling points. There is also a freezing point. When water is cooled enough, it starts to slow everything down. The water molecules eventually mellow out so much that they're able to start locking

into place. They arrange themselves, forming organized little rings. It's such a standardized, repeating positioning that ice is technically a crystal, which means, that to a geologist, ice could be considered a rock.

Water molecules in ice are arranged in little hexagons, and the best part is that this hexagonal arrangement eventually turns into something we can actually see. Up close, snowflakes have six spokes because of the six-sided polygons that the water molecules form. While we can't see the individual atoms and molecules, we can still get a glimpse of what they're up to even on our wildly different scale of existence.

If you don't live somewhere where it routinely gets cold enough for spontaneous snowflakes to fall from the skies, you can sometimes find hexagonal ice crystals on long-forgotten freezer-burned leftovers. It's the literal silver lining of those ruined enchiladas.

Thanks, Water!: How Water Lives Up to the Hype

There's a very good reason we've based our perception of the world around water: this substance has made life on Earth possible. We're the only planet in our solar system with so much of the stuff on its surface, partly because, thankfully, our planet managed to hang on to it for this long. Our atmosphere's air pressure doesn't just make it hard to boil pasta water; it also keeps our planet's water from floating out into space. And we owe some of this to Earth's magnetic field, which shields us from particles that would otherwise strip away our atmosphere and then rid us of water too. Every time you take a sip of water, you should once again thank the Earth's iron-rich, churning core for helping preserve our water resources so that our species could evolve and continue to live here.

Strangely, we don't know exactly how we wound up with so much of this wet stuff. It seems as though part of it came from ancient volcanic eruptions, liberating it from some of our planet's early guts—the space chunks that eventually coalesced into our planet when the solar system formed. But we may have gotten some water since then, in the form of ice hitching a ride on meteors and comets, crashing into Earth with the most obnoxious water delivery system of all time.

But however water got to Earth, it has served living things incredibly well over these past billions of years. Even a little bit of dampness can make it possible for tiny organisms to survive, which is why I have to clean my shower so often. Mildew fungus likes water just as much as I do. Where there is water, there is life. Even if I'd sometimes prefer it not to be there.

But what is so special about this otherwise simple molecule? Couldn't something else be as useful? Well, actually, no. Water lives up to the hype. It's got some great qualities.

One of water's best features is its resistance to change. But not, like, good change like gender-neutral bathrooms. Water resists *temperature* changes much better than other substances.

It takes a long time to boil water, which frustrates me when I'm waiting for dinner, because water can absorb a fair bit of heat without changing temperature. This goes back to water "liking" itself so much, as we talked about earlier in this chapter. Those small attractive forces have to be overcome for water to change from a liquid to a gas. Many other substances require far less energy to heat up, which is why even on a scorching hot day, a pool of water can stay refreshingly cool while the concrete or tile or whatever is around the pool can be blisteringly hot. And because our bodies are made mostly of water, this feature helps us resist temperature changes too.

So water can take a lot of heat, but when it *loses* a lot of heat it does something interesting too. It changes to a solid, which we call ice. That's not the interesting part, though. All substances have a temperature where they will turn into a solid—even things we think of as gases like oxygen, nitrogen, and carbon dioxide. But ice does something rare: it floats.

Most solid substances are denser than their liquid form. If you dropped a chunk of solid silver in liquid silver, it would sink to the bottom. But water ice crystals' hexagonal structure has so much empty space in it that it becomes *less* dense than liquid water—a fantastic outcome. If ice sank, whole lakes or even oceans could freeze solid when the temperature dropped below water's freezing point. But since ice floats it enables the water below it to stay liquid.

Lastly, and most fortuitously, water is great at dissolving things. Since it has positive and negative ends, pretty much anything that has a slight charge will dissolve in it. Things like salts, sugars, acids—they all happily dissolve in water. In our bodies, water is the bath that all these materials float in to get to their destinations and keep the machine running. And just as important as the things that dissolve in water are the fats and oils that don't. The edges of our cells, which are tiny water sacs, are surrounded by fatty membranes. Because they can't be dissolved in water, they form a convenient barrier.

If water were overrated, I'd be honest with you about that. But it really is impressive: it resists temperature changes, its solid floats on top of its liquid, and it dissolves lots of things. These handy qualities are, again, the only reason any living things are here on Earth. And based on how well water has served us here, its presence is the best predictor we know of for finding life on other planets. But if we ever do discover extraterrestrial life that doesn't use water, I'll be interested to hear how they managed to do it.

Water, Water, Seriously Everywhere: The Everyday Water Cycle

There is water surrounding us, even when we're not poolside or taking a shower. It's in the very air we breathe. Sometimes we're aware of this when it's particularly humid, and you can feel the moisture in the air, but other times it's easy to forget about the water that is around us all the time. We inhale it with every breath. The water in our sweat slowly evaporates from our skin. But it's not just us. Other animals are giving off water. And plants lose much of their water to the surrounding air as the day drags on.

It might hang around for a while, happy to share our space with us. If you have a glass of ice water, some of the water molecules in the air will collide with the cold surface and change back into liquid, forming the beads of water on the side of the glass, dripping down onto the coaster you surely put on your coffee table, right? If not, go get one right now, you animal.

Some of the water will float up, up, up into the sky where it can make new friends and form a cloud. Whether it's a streaky cirrus cloud, a puffy nimbus, or a foreboding dark storm cloud, that's all water up there. It floats over us, looking down and probably judging us.

When a cloud gets so waterlogged that it can't stay aloft anymore, that water falls down to us, and depending on the temperature, it might be liquid or frozen crystals of water (rain or snow). Just like temperature, weather events like these impact our daily lives, but I'm ordinarily more concerned with their effect on traffic conditions and my plans to be outside than I am about appreciating what is happening in the sky. But if I don't have any distractions, sky water is one of my favorite things, although mostly because I've spent so much time living in deserts where it's a rare treat.

Earth is the only planet in our solar system that has sky water in these quantities, but that doesn't mean no other places have clouds or get rain. Venus's sulfuric acid clouds drizzle down onto the planet. Saturn's moon, Titan, has methane showers. And Mars has carbon dioxide clouds sometimes.

Every time you breathe in air or see a cloud on your drive to work, you're experiencing the real effects of the nature of water on our planet as it moves from place to place. And more importantly, the tendency for water to move in all these wondrous and various ways is the only reason we are here.

What's in *Your* Water?: There's More Than Just Water in There

After all this talk about water, H_2O, dihydrogen monoxide, we probably ought to discuss what else is in a typical glass of water, because it's more than water molecules. Well, unless you are drinking expertly distilled water on a regular basis. The rest of us are drinking tap water, filtered tap water, or some sort of packaged water (which is likely filtered tap water). And in that case, there are other atoms in there too. Even a few living things here and there.

First you have some salts and metals. You'll find sodium chloride, our old friend table salt, as well as things like calcium carbonate, which leave those hard water stains. There are also things like iron from Earth's crust. Yes, your water has traces of Earth guts in it because that is, in fact, where it came from. Not, as you might otherwise think, a magical water factory. It also may have fluoride, which many municipal water departments add for its proven benefits to our tooth enamel.

There are some bacteria in there, too, but not usually in any amount that it's a concern. There may also be the occasional amoeba. These actually can be of concern, but only if you put the water way up your nose, which doesn't normally happen unless you are using one of those sinus-cleansing pots. And that is why

the directions instruct you to boil the water (and let it cool down!) before you use it. Amoebas can't survive a pot of boiling water.

But apart from a brain-eating amoeba here and there, your water is by and large made up of nonliving materials. And even the scarier-sounding contaminants of water (like arsenic, uranium, and hexavalent chromium of *Erin Brockovich* fame) usually occur in such minuscule amounts that they're not anything you need to worry about on a daily basis.

Another common concern is lead, which can leach from pipes. Now, let me quickly mention that I often like to tell people not to worry about things. If you compiled all my texts and emails, my most oft-used phrase might in fact be "No worries." But you can't *hakuna matata* your way out of this one. Lead is *bad*. This metal replaces or interferes with crucial elements in our bodies like calcium, iron, and zinc, messing up a long list of bodily functions, and to make things worse, it's very hard for our bodies to get rid of it. The semi-good news is that if there is a significant lead problem with your tap water, you probably know about it already, and you can look up information about your city's water online if you have any concerns.

But I should also mention that you should be very wary of googling questions about tap water, distilled water, and water contaminants. You will find some very suspicious sources high up on the search results. It would seem as though water is not just vital to our bodies (and all living things), it is also the lifeblood of conspiracy theorists and anarchist wackadoos. Because few things are as disturbing as the thought of invisible mind-controlling contaminants in your water, added by a fiendish supergovernment. And no one knows this better than people who are selling very expensive water filtration systems. Be extra careful where you are sourcing your water information, not just your water. And if someone is selling you fear, how much are you paying them? Take me out to dinner instead.

Quench Your Thirst: Why We Need to Stay Hydrated

On a hot day after working up a sweat (whether by a vigorous jog or taking the garbage out to the curb), almost nothing beats a cold glass of water with a few cubes of ice clinking around. Since our bodies need so much water, we come equipped with a built-in water meter that activates a sensation we call thirst, the deep desire to gulp down delicious dihydrogen monoxide.

Just like I'm surprised I can't be outside for even an hour without sustaining a sunburn, I'm sometimes frustrated by the frequency with which I must ingest water. It seems like an unreasonable requirement. Several times a day I simply have to stop what I'm doing and drink. Fish never have to think about it at all, those lucky jerks. They can conduct all their fish business without water bottles. But the annoyance of having to bring water with me if I'm going to be away from a faucet for several hours is nothing compared to the downsides of even minor dehydration. Without the right amount of water in our bodies, we can get headaches, feel irritable, or even lose consciousness.

Similar to how any ecosystem needs a source of water, so do our bodies. Each of us is our own little island, and we have to manage our water intake carefully. Because getting dehydrated is not just unhealthy, it can be downright fatal.

When thirst sets in, our bodies try to conserve what water we already have. Our body's filtration system uses less of it in our urine, which is why when you're dehydrated your pee is darker and often therefore smellier. And if you let dehydration continue, you'll even start to sweat less. While that might sound nice for those of us who sweat too much, this is not ideal because it means your body can't cool off as effectively.

If you insist on forgoing water even longer, the mighty blood rivers of our little island start to get a little drier, which means they don't flow as well. If this continues, the body takes some extreme measures to try to keep you alive, like slowing blood flow to organs such as your liver and kidneys. That is often the eventual cause of death in a dehydration scenario, either liver or kidney failure.

But apart from desperate survival situations, this need not happen to you. Stop what you're doing right now and have a sip or two of water. And as you enjoy it, think about how those water molecules have been here on Earth for billions of years. They used to be in the ocean. They were once flying high above the Earth's surface in a cloud. They could have once been in the bladder of a dinosaur. Every sip of water brings you up close to the history of our planet and the living things on it. Drinking this wonder substance not only saves you from a grisly death, it's an act that links you to every living thing that has ever called our planet home. We all need this substance, and when we've finished with it, breathing it out, peeing it out, sweating it out, or any other form of excretion, it reenters this cycle older than life itself, to sustain and bathe the next thing. Or to sit in the ground and be alone because, hey, it's not just here for our amusement.

Would you like some water? It's a rare vintage, many billions of years old...
MENU
MENU

CHAPTER 7

Small but Mighty: The Many Lives of Cells, from Stand-Alone Creatures to Multicellular Wonders Like Us

To see something truly awe-inspiring, you need only look in a mirror. You're one person, but you're also trillions of cells. Your cells do everything for you, allowing you to move through the world, feel it, make sense of it, and read about it in fantastic books. They're your organs and tissues, the blood and guts of your body. You don't see them much if things are going well, so you can easily forget about what they're doing and how hard they work for you, the sweet little things.

And as remarkable as they are, you are even more than your body cells. Part of Planet You are communities of microscopic inhabitants (bacteria, for the most part) that set up shop in every nook and cranny of your warm, damp, oily body. You are, in fact, quite outnumbered by them. Hardly as invasive as it sounds, these microbes help make us who we are, and understanding the complex relationships between our cells and their microbial neighbors is the next frontier of health and medicine.

And that's just our bodies. We also live in a microbial world. Every surface, breath of air, and sip of water has scores of cells too small for us to see. Sometimes they eat our food, occasionally they make us sick, but often they're lovely neighbors. In this chapter, we'll spend some time getting to know them.

Your Start As One Cell: We All Have This Humble Beginning

I don't know how old you are. Maybe you're still growing taller, or perhaps your growth spurt is a distant memory. But isn't it truly incredible that at the very beginning of your journey through life, you were but one little teeny tiny cell? Just *one*. You were, oh my word, so small.

You were adorable!

(what happened?)

We don't have to back up too much further than that, because odds are you don't want to think at length about your mother's egg, your father's sperm, and um, how they wound up meeting...but still, that they did, and years later, here you are.

Astonishingly, you went from exactly one cell to trillions by the time you were born. That's quite the pace. When I was pregnant and picturing the size of the developing fetus that would eventually turn into a human baby, it was unnerving to think about how big she was so early in the process. I remember when she was the size of a lentil—or so the pregnancy charts told me—even that seemed humongous. I held my thumb and forefinger together, with a lentil-sized gap, and thought, "Wow, that's in there."

My daughter's cells—and years ago, *your* cells—were growing, splitting in two, growing some more, and splitting again. This system of dividing in half is the most ancient form of growth and reproduction. Bacteria do it. Dinosaurs did it. Our cells do it. We all do it. No one has thought of a better way to get this job done. Grow. Divide. Grow. Divide. Grow some more. Divide some more. It never gets old (although we do).

You, right now reading this book, are the product of many years of cell division, with your cells now numbering about thirty trillion. And no matter how old you are, it's still going on. Each and every day, your cells decide on their own schedule that it's time to make another one of themselves. Right now it's happening, but not in equal amounts everywhere in your body. Some cells stop dividing once they've established themselves. Your heart, for instance, beating away in your rib cage, is made mostly of muscle cells that don't divide. That's also why you've probably never heard of someone getting heart cancer—there isn't much cell division happening there, so the odds of uncontrolled cell growth are slim.

Your skin, though, is a place of constant regeneration, and it's one of the organs we can actually see. Well, you can see the dead cells on top, at least. Underneath there are cells dividing and dividing over and over again to replace these layers.

We shed these cells constantly, contributing to small piles of dust around our homes. Yet, even after losing countless cells every time I sashay through a room, I continue being me without them.

We think of ourselves as a unit. I am me. Look at me. Here I stand being me. But what am I, really? I'm a teeming community of individual cells that all emanated from just one. So are you. So is everyone you have ever met. We're meandering cellular metropolises that all began life the same way, no matter where we are now. We all have that in common.

Your Cells Now: The Types of Cells in Your Body

Whether they're dividing constantly or have been with you for nearly your whole life, do you ever stop to think about your assortment of cells? They're all you, yet they're unique.

Your thirty trillion or so cells come in a wide variety of styles. You have skin cells, blood cells, muscle cells, immune cells, nerve cells, fat cells, and so many more that listing them would be ridiculous. There are many, many hundreds of different types of cells making up your body right now, and they're all doing a pretty decent job being the many pieces of you.

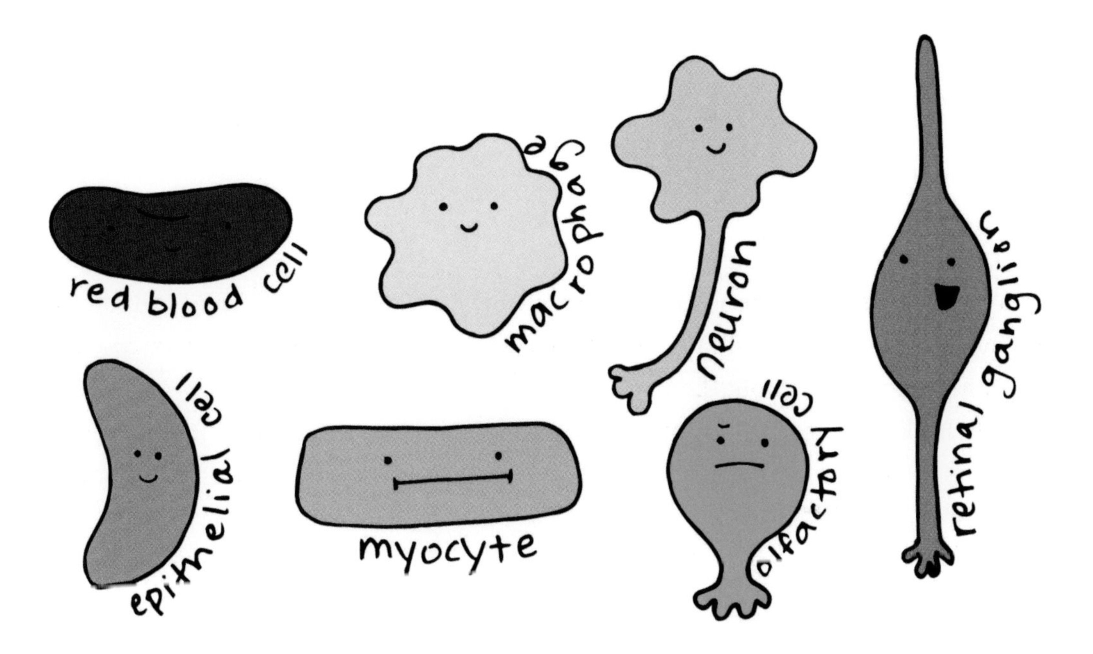

The muscle cells in your heart twitch and contract in rhythm to work the pump that pushes red blood cells—another distinct type of cell—around your body, bringing oxygen to yet more types of cells to keep their engines humming. Your liver cells filter your blood. Your lung cells exchange gases. They all work tirelessly to keep the giant machine that is you running.

In yoga classes and mindfulness meditation workshops, you're told to listen to your breath, focusing on your lungs with each inhale and exhale. But try closing your eyes (which I guess means you have to stop reading for a moment here) and thinking about each of your distinct organs and tissues and the cells within them that are doing their various jobs. Feel your heart beating. Try to "listen" to your brain, if you can wrap your head around it. Check in with your colon. Have a chat with your liver.

Most of us think about our bodies only when there's trouble. When I'm sick, I'm viscerally aware of every ache, pain, and bodily annoyance, but as soon as it's over, I resume the status quo, neglecting my body and taking it for granted, like I do so much else.

Even with all their annoyances (because who doesn't have a few?) our bodies are truly spectacular, and we still don't totally understand them. After thousands of years of poking and prodding, we continue to uncover the complexities of our cells, tissues, and organs and how they interact with one another. The prevailing medical wisdom today is different than it was a decade ago, and if we could fast-forward a dozen years, just imagine what new treatments we might embrace for depression, cancer, or allergies. We don't have all this business figured out yet. And while that's definitely frustrating for those of us suffering now, it's also inspiring that we're all so complicated and unique. There is a lot left to discover within ourselves, so revel in the mystery that is your body.

Ouch!: How Your Body Repairs Damage

I don't think I could be a surgeon or anything like that, but I'm not terribly squeamish. I'm slightly the opposite in fact. When I cut myself—you know, a normal, clumsily-cutting-an-apple-type slip of the knife—I take a moment to stop and behold the dripping blood. *Look!* I think, *Actual living cells! What a pretty shade of red!*

So much of what you see on the outside of your body is just dead stuff. Our skin is layer upon layer of these old deceased cells protecting the living tissue below. That's why you don't bleed when you lightly scratch an itch or rub a towel over your face. And hair isn't alive or dead. It's long strands of proteins.

We're not accustomed to seeing much of our living tissues, so when we do come across them, any sense of wonder is understandably coated in a thick layer of alarm. No one plans on meeting these inward cells, even though they are so very nice.

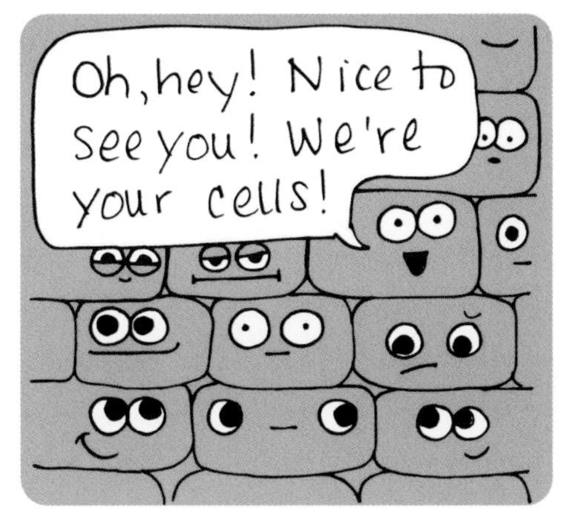

On the hopefully rare occasions when you injure yourself, your body immediately sets off a chain of events to repair the damage. Let's think about a simple paper cut. I get those more often than I'd care to admit. Something must be deeply wrong with me, and I should probably invest in a letter opener or only handle paper while wearing oven mitts.

Anyway, when you slice down into your skin, the cells nearby definitely notice. First of all, you killed a few of them. Oops. They are the immediate casualties, but that's okay, you'll make some more. The spilled guts of the fallen cells contain molecules that trigger nearby immune cells to protect the vulnerable opening from any invaders, which also causes blood to flood the area. It all gets very crowded and chaotic.

Then you scramble some platelets, which cause the blood to clump up and plug the hole in the dam that is your skin. Then, once you stop bleeding, the cleanup and regrowth can begin. The surviving neighbor cells start to divide, eager to bridge the gap in the wall. Once they meet in the middle, though, this flurry of cell division stops.

The whole process is similar to a structure fire in a city (one that has a well-funded fire department and strong unions). The alarms go off, fire engines show up to stop the damage, and an ambulance helps anyone affected. Neighbors come outside with bottles of water and granola bars. And then slowly the house gets rebuilt. It might seem frantic at times, but it's a tactical exercise with standardized procedures.

The next time you get a little cut or jab, after the pain subsides, you can marvel at the cells that are at that moment kicking into gear to protect you and fix the damage you've done.

And then, be more careful next time.

Your Microbiome: The Cells in Your Body That Aren't Quite *You*

Take stock of your body. Does it all feel like your own? I know it might seem that way. But it's not all *you,* technically. There are trillions upon trillions of bacteria on and inside you.

It's weird to look in the mirror and ponder this. We all have our occasional identity crises, but this one is a doozy. Generally speaking, the bacteria of your body, as far as individual cells go, outnumber cells that are *yours.* You are made of about thirty trillion cells, but you've been colonized by as many as thirty-nine trillion bacteria. You're more bacteria than human on a cell-by-cell basis. Is your body even your own?!

But hold on, even though the bacteria are more numerous, they do not outweigh us. Since each bacterium is so many times smaller than one of our cells, the bulk of you, I'm happy to report, is still *you.*

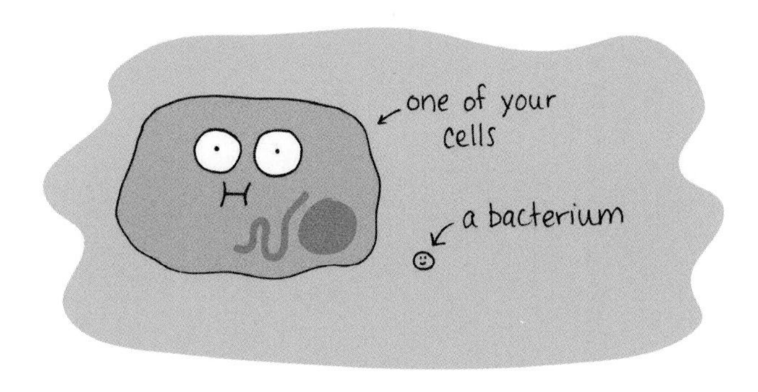

But matter isn't all that matters. The microbes that share our real estate have an outsized impact on us. In some ways, they make you who you are as much as your actual cells do. Healthy communities of bacteria keep harmful invaders from claiming space in your body. They help you digest your food and draw nourishment from it. They calibrate your immune system, teaching it the difference between harmless allergens and dangerous intruders. And far from being mere squatters in your body, bacteria are instead loyal tenants who help with improvements: they can tell your own cells which genes to use, and sometimes provide you with genes too. That blurs this imaginary line between "us" and "them," and it makes you wonder who is truly in charge around here.

Our relationship with these microscopic beings means that if you had spent your life in a sterile environment (perhaps in a squeaky-clean plastic bubble) you might not be the same person you are today. If you ever picture alternate timelines of your life—if you had grown up in a different time or place, if you had made different decisions—make sure you include this version. Without microbes, or with exposure to different strains of them, you could be slightly different. Whether you find that disturbing or intriguing, I hope it makes you feel even more special than you did a few seconds ago.

We're still figuring out the intricacies of these relationships, and we don't yet know the identities of all the bacteria that call us home and affect so much of our health. So that means inside you right now could be bacterial strains new to science that no one knows anything about yet. Your innards are like the Amazon rainforest, full of undiscovered biodiversity and unknown ecosystems. You could have a totally different microbiome than the person sitting next to you. Or a person who lived two hundred years ago.

And you're not exactly the same from one day to the next either. Your dietary decisions and the other things you put in and on your body can affect your teeming microbiome. The ones that live in our digestive tract have no say in their diet; they only get to eat the food we choose to shove in our faces. If you ever need inspiration

for eating "better" (whatever that means to you), you can picture the trillions of bacteria that are counting on you for nourishment.

If you're a busy person, this might sound like yet another thing you have to manage—and I know, ugh. It doesn't seem right that microscopic creatures could affect how we feel on a day-to-day basis, but that does indeed seem to be the case. But when your microbiome is thriving, so can you. So even from an entirely selfish standpoint, it's best to look out for these little guys.

On the surface, though, our microbial cohabitants can make us stinky. Have you ever had bad breath or body odor? No? Well, congratulations to you. But if you're like me, the stench wafting from your face hole or underarms is largely due to bacteria. They feast on the food we eat or the oils and proteins we make in our sweat glands, and their waste products, well, they smell pretty bad to us. If you feel uncomfortable with these odors, you can take some solace in the fact that it's not quite *you* that makes them. But either way, a shower and a good toothbrush will solve the problem in the short term.

Bacterial Murderers: The Ins and Outs of Antibiotics

It is truly astounding that if you go to the doctor with an ear infection, skin infection, upper respiratory infection—you know, any of these bacterial communities run amok—your doctor can prescribe you something that for a long time was just the stuff of our wildest dreams: a substance that will lay waste to the bacteria but leave your cells alone.

We think nothing of this, but for people a few generations back, these infections could easily be fatal. An ear infection could spread to your brain and be the end for you. For people in the developed world today, thanks to medicines we should be very grateful for, dying from a simple infection is now a relatively rare occurrence (and hopefully it stays that way).

Think of the last time you took antibiotics or assisted someone else with them. My last brush with these compounds was when my daughter had an ear infection. She was prescribed penicillin, the very first type of antibiotic we humans discovered and produced on an industrial scale to treat infections. Although, I'm not a big fan of the word *discovered* in this instance. It sounds very cheerleader, like "Rah rah rah! Go humans! We're the best!" But it's not like we invented penicillin with our own ingenuity and engineering prowess. I think a better term for the discovery of antibiotics would be that we *noticed* them. Penicillin is what some fungi excrete to fend off bacteria, and we frankly just stole their idea.

Penicillin makes it hard for bacteria to build their cell walls. In fact, many antibiotics work this way. They attack structures that our body cells don't have so the drug can destroy only the bacteria and leave *our* cells unharmed. Every time you've taken antibiotics, you've been ingesting a drug that selectively murders only some kinds of cells.

However, these drugs aren't so scrupulous that they kill only the types of bacteria that are causing you trouble. After all that talk about our microbiomes and how we should strive to care for them, we then go and take drugs that lay waste to a good chunk of them. Antibiotics can't differentiate between intruding bacteria and the ones that we want to stick around, also known as "bad" and "good" bacteria. I hesitate to even mention this common judgment because it's subjective, like many of the

topics we've talked about so far. Nothing in nature is inherently bad. It's never that simple. Just like weeds aren't bad plants, bacteria, even when they're about to kill us, are merely misplaced cells growing successfully in a place we'd really rather they didn't. But it's not like they're doing it on purpose.

Antibiotics are such a miraculous medical treatment, we can easily be lulled into thinking they will always be there for us—that they'll forever kill the bacteria that would otherwise harm our body cells. But bacteria were never consulted on this, and no treaty has been signed. They're fighting back and developing resistance to many of the common antibiotics we use. That's scary, but it's also pretty impressive.

These are living things, after all, and in all the *Jurassic Park* movies we are reminded that life is always trying to break free of any restraints we deign to put on it. Bacteria are no different. Even I can forget that these are small living things causing these troubles for us, whether it's a minor sinus infection or something more serious like tuberculosis, tetanus, or diphtheria. They find ways around the roadblocks that antibiotics set up for them, and what's ingenious (or fiendish, depending on how you look at it), is that they have a system for sharing information with one another.

From their perspective, *we* are the villains. In Star Wars terms, we're the Empire; they're the Rebellion. So at the end of the day, rather than demonizing bacteria, we should remind ourselves that we don't live in a sterile world of all-or-nothing absolutes where living things are either benevolent or evil. We share our world with these creatures, and they have just as much claim to it as we do. And if single-celled creatures can outsmart us so easily, maybe we should tone down our egos a bit.

Cells Everywhere: We Live in a Microbial World

Pondering the eternal struggle between good and evil has made me hungry, so I'm going to run to the fridge to grab a snack. But before I do that, let's consider why we all have this obligatory giant cold box. To sum it up, it's because bacteria, like most of us, don't like the cold.

Yes, from the moment we bake a loaf of bread, pick a strawberry from the plant, or pull a piece of chicken off the grill, we are in a race to see who can eat that food first—us or the many microbes that share our world.

When microbes get to our food first (without our consent, as things like beer and cheese *require* microbes), we say the food has "gone bad" or "spoiled." But really it means we lost the race. There isn't a prize for second place either. It's all or nothing, this competition.

But now, what are the microbes exactly? When it comes to food spoilage, the big ones are bacteria (which you probably guessed) and fungus. Fungus is my personal favorite because you can so often see it with the naked eye. I can tell when my bread has been compromised by fungus because bread mold is very apparent, especially the green kinds. The white ones are sometimes harder to notice at first because they look like a light dusting of flour—a normal occurrence on some rustic breads. And I must confess that on some of my more absentminded days, I have been known to bite into a piece of moldy bread. But even then, I notice pretty quickly. If I'm too busy to get the visual cues, the smell is quite helpful. And by helpful I mean awful.

One of my most persistent quandaries is that Wednesday night question: are these weekend leftovers safe to eat? It bothers me to no end (seriously to *no* end) that leftovers are not suitable to eat for the same length of time as a standard workweek. I would like to cook a bunch of food on the weekend and eat it until Friday, but that usually doesn't work. But when we wonder about those leftovers, what we're asking is, chillingly, *has anything else started to eat these*?

We're lucky that bacteria, as resourceful as they are, these little creatures that colonized Earth so long ago and have been wildly successful in every conceivable environment, don't like eating our food when it's in the fridge. Even for hardy bacteria, there is no getting around the fact that near water's freezing point (most fridges are just a few degrees above this threshold), things move a little more slowly. Enzymes, which are proteins that help reactions happen within cells, don't work as effectively when the temperature drops. Inside a refrigerator, everything slows to a glacial pace. It buys you and me enough time to finish the meatloaf, down the raspberries, and toss back the milk.

However, if food is left out at room temperature for even just a few hours, the bacteria that are around us all the time can colonize the food, multiply, build vast bacterial cities, and draft a constitution. Yeah, fine, not that last thing, but at temperatures they find more comfortable (which are in the temperature ranges we enjoy too) bacteria can start to eat our food.

That wouldn't be so bad, maybe, if they were like people when they had to go to the bathroom. If they got up, left the food behind, and deposited their excrement somewhere else, maybe we could all get along. But no, while bacteria aren't "bad," as we talked about in the previous section, they are not always the best housekeepers. They literally poop where they eat.

Luckily for us, we can smell some of the waste products they make. When we give questionable food a sniff before eating it, we're trying to detect bacteria poop. (But honestly, if you're going this far, maybe it's best to just toss it.)

And while smelling bacterial excrement is pretty gross, the worst situation of all is eating viable, thriving colonies of certain types of bacteria that continue flourishing in our bodies. Introducing troublesome newcomers in significant numbers sets off some alarms, and our bodies are prepared to do whatever it takes to restore the balance. We might insist that these questionable items leave the way they came in (vomiting), or we flood the area with water to wash them away (diarrhea).

This is why public health and common courtesy stress the need to wash your hands in general but especially before preparing food, because a significant amount of bacteria on your hands could then be transferred to that otherwise delicious pizza, where the bacteria could happily multiply. Food poisoning is unpleasant enough without having to fully reflect upon the fact that basically it means someone who handled your food didn't wash their hands well enough after wiping their nether regions. (Although it's also possible someone sneezed all over your tacos, so it's not all about fecal bacteria.)

Yes, there are bacteria all over the place. We're clearly aware of this because we use soap, wipe down countertops, and put food in the fridge. But you have somehow survived this onslaught of microbes long enough to reach this chapter, not to mention the entirety of your life before you picked up this book (and really started living, am I right?), so you're probably going to be okay. You'll just need to keep doing what you've been doing—putting milk back in the refrigerator, not dunking your cookies in the toilet, and for the love of Pete washing your hands before you make lunch.

Tiny Deaths: When Our Cells Die but We Don't

The more morbid among us like to muse that we are all slowly dying from the moment we're born. I suppose that's arguably true. I think it makes slightly more sense to say you start dying when you stop growing. That means I've been decaying since I was about thirteen. Based on a cursory evaluation of life since middle school, I'd say that was *mostly* inaccurate. But indeed, parts of us are dying every day. Right now, somewhere in your body, a cell is deciding that it's time to go. I know, it sounds so dramatic. But it's a normal, dignified process.

See, your cells (unlike most cable TV shows) know when to call it quits. This is completely different from being killed by a paper cut or a viral infection or other trauma. No, this is planned cell death, which is calm, orderly, structured. It's all part of the grand plan, and when cells get old, death doesn't catch them by surprise.

It's a bit like a business deciding to close. It breaks down all its equipment, lets its neighbors know, and makes the space available for the next tenant. Where this analogy breaks down, though, is how the dying cell meets its ultimate end—it gets eaten.

Yes, you have cells tasked with going around and gobbling up your dying cells. These hungry cells are called macrophages, and they're part of your immune system. They also eat invaders, so it's not all about you, but think about this: somewhere inside you, let's say your pancreas, right this moment a cell is closing for business and letting any passing macrophages know it's time for a snack. The macrophage happily obliges, enveloping the old-timer, breaking it down, and recycling all its parts so other cells can use them. This keeps everything nice and clean, allowing nearby cells to keep humming along helping you live. These macrophages literally save us from ourselves.

Again, you might feel like one unit, but you're really an island teeming with life. And like any diverse, thriving ecosystem, you're built on birth, death, and a few things in between. In my case, as a fully grown adult, there are roughly equal amounts of birth and death going on. As my cells die, other cells split in two to replace them, so I stay about the same size and keep on breathing.

You started out as one cell and grew to be who you are today by dividing over and over and over (and over and over). While you revel in the living cells you have in this moment, don't forget that you equally owe your current state to the cells who died to get you here. It's not that unlike your family ancestry. You are the culmination of generations of cells and generations of people, on different time and size scales.

But don't be sad about your fallen cells. There's no need to hold a funeral for them; instead you can have a celebration of life, experiencing today to the best of your ability. It's what your dearly departed cells would have wanted.

CHAPTER 8

Our Squishy Brain Blob: Mind-Blowing Neuroscience

I'm using my brain to think about brains. I could probably end the chapter here, because that's completely astonishing to me. But, for good measure, let's keep going down this mental rabbit hole.

The study of the brain and the rest of our nervous system remains one of the big frontiers of science. There is much we still have to learn about how this annoyingly complex system works. And not just because it's nice to know. We need to help people with mental illnesses like depression and chronic conditions like epilepsy, as well as degenerative diseases like Alzheimer's. In the past, a common way to learn about the human brain was to note what happened when someone had an injury. By seeing the effects of damaging certain portions of the brain, we could infer what its purpose was when it was intact. We can still learn that way, but now we have some more sophisticated ways of doing this, with simulations, models, artificial intelligence, and a suite of imaging tools like MRI (magnetic resonance imaging), which help us map the brain and try to understand its interconnectedness.

But the many little things our brains do each and every day can be just as mind-blowing as the most recent neuroscientific revelations. Even the seemingly simple act of reading a book involves processing visual stimuli (or auditory if someone were to read it to you), language decoding, and memory. And you also have motor functions that are helping you sit up, turn the page, and perhaps occasionally take a sip of coffee or bite of bagel. (That last part is what I'm doing right now.)

Our brains define us, and they allow us to make sense of our (very confusing) world. The brain is always working, even when we're sound asleep. But somehow what we hardly ever do with this organ is use it to reflect upon its own magnificence. We easily ignore the very organ we use to focus on things. Our brains deserve extra attention, so let's give them some time in the limelight and take a look at how our brains allow us to see, remember, and sometimes stress out.

Look!: How Our Brains Attempt to Make Sense of Our Surroundings

For starters, nothing is more meta and mind-bending than thinking about how your brain perceives your surroundings. Remember, you are merely a squishy flesh suit propped up by a hard skeleton, moving through this strange world, receiving information from it and trying to make sense of it with each passing moment. It should be exhausting to merely exist, but our brains so excel at navigating our world, we don't even have to "think" about it.

Even when you're not doing much (which is my favorite thing to do), your brain is crunching enormous amounts of data, the heaps of information coming from your eyes, ears, skin, nose, and tongue. But your brain isn't processing it all equally. It ignores the vast majority, but even then it still has to decide what is worth paying attention to and what is inconsequential enough to disregard.

Think about all the information you're receiving right now that you can blissfully ignore: perhaps some ambient sounds that fade into the background, the physical sensation of your butt resting on a chair, a poster so old you hardly realize it's there, or maybe the slight stench of the dirty dishes you don't feel like doing yet.

Sometimes we disregard things because our nerve cells stop bothering to tell us about them. When you put on pants in the morning, you probably felt the fabric brush against your skin. If you have reasonably comfortable pants, though, you're not thinking about them touching your skin every minute of the day. You eventually get used to them, partly because the nerves in your skin get tired of sending the same signal to your brain over and over again. It's called adaptation. It's how your nerve cells manage to get on with their day.

Interestingly, the same thing can happen to our vision. If your eyes were completely motionless and you stared at the *Mona Lisa* for a while, it would disappear in front of you, as the nerves would become used to that image and commence ignoring it. However, this doesn't happen even if you stare at that painting for an hour (like some people do) because our eyes are never completely still. They're always jutting back and forth, even just slightly. It's incredibly hard to stare at one point intensely for even ten seconds. I mean, that's why they call them staring "contests" and not staring "vacations."

But if you dare (and don't have any vision problems you could exacerbate), you can experience what it's like for the nerve cells in your retina to stop sending information about something you're looking at. Pick a spot to stare at, and then hold your eyes in place with your fingers. What I find easiest is to put my thumbs on my bottom eyelids and my forefingers on my top eyelids and gently press the outer edges of my eyeballs so they are completely still. Now continue staring. Within a few seconds, your peripheral vision will start to go dark, and then the rest of your field of vision will wipe away as well. The nerve cells, tired of sending the exact same information to the brain,

eventually stop saying anything at all. For me, it takes only ten seconds of completely motionless staring for my vision to go completely blank.

But do not despair! Your vision is easily returned when you let go of your eyes, look around, and blink. The nerve cells, now with new visual stimuli to tell the brain about, resume transmitting information.

The way our nerve cells work is similar to how journalists at a newspaper work. They only report on things that are new and different, and won't tell you about things that are staying the same. It's not front-page news that the sun has come up for the bazillionth time. No one is going to report on that. But if the sun were to come up wearing sunglasses, we'd definitely hear about it incessantly across all news platforms.

Even when your eyes are happily relaying the freshest visual gossip to your brain, they still have a blind spot. In the back of your eye where you have all your light-sensing nerves that allow you to see, there is a gap where the optic nerve plugs into your eye. Luckily it's not in the middle of your retina; it's off to the side, in a less vital part of your field of vision. But still, when you look outward, there should be two black spots on either side of your lower periphery because there are no nerve cells to process that part of your view. But obviously you don't see two dark patches following you wherever you go. Your brain is filling in those gaps in real time. Yes, a portion of what you see every minute of every day is not actually there, but is your brain's best guess as to what probably belongs in that spot. We're all slightly hallucinating all day long.

But wow, it makes such a good guess. It sure seems like I can see things as they truly are throughout my entire field of vision. I wish my brain were this good at guessing in all respects, so I could be more useful at trivia nights.

If you want to see proof of this blind spot, here is a quick optical test to expose this mental trickery. Close your right eye, and stare at the plus sign below. If you move the book back and forth, you will eventually see the star disappear, as it has lined up with your blind spot. Then try the opposite, close your left eye, and stare at the star, and you can make the plus sign disappear.

We put a lot of stock in our senses, claiming that "seeing is believing," but it's the opposite. We have to believe what we see to get through the day. After all, this is all we've got to make sense of the visual world. But our vision is not a camera, giving equal attention to everything in the frame and capturing an objective view of the world. Our vision is selective, partially blind, and easily manipulated. I hope you can now *see* what I mean.

Misremembering: We Forget That Our Memory Isn't So Accurate

Not only does our brain do some heavy lifting to process the things we see, hear, feel, and taste in real time, it also stores some of that information for later. Let's take a trip down the brain's Memory Lane.

This might get a tad awkward, because nothing is more personal than our memories. Each of ours is unique. No one else on Earth has the set of memories you have. Even someone who follows you everywhere would have different ones. And you've spent years making them. It's a lot of work. Life is the practice of making memories, day after day and year after year.

Living with a two-year-old, I'm watching memories get filed away every day. She's learning things one by one. *Ice cream is delicious. Apparently, I'm not supposed to put my shoes in my mouth. Screaming, when wielded appropriately, is an effective tool to gain attention.*

Our memories, even the fleeting ones, form much of our identity, and we depend on them completely to get through our world, but sometimes we can rely on them so much that we manage to "forget" to appreciate them. You see, memories are this abstract idea, but their existence depends on physical things in our bodies. They don't just come from the ether, even though it seems that way when we strain to remember something and it pops into our head hours later.

Think of a fond memory. Maybe a vacation, get-together, or favorite movie. Ponder what is happening right now in your brain as you search for that memory and relive it. Brain cells are chattering, sending messages to one another, and pulling sensory information from that saved experience. Maybe you can see some of the details your eyes processed that day. Maybe you can smell, hear, or feel what you did in the past.

Thinking about memories can be a mental workout after a while. They seem so real and tangible to me, like there is a film reel in my head, with moments I can queue up. But my memory theater has only one seat. No one else can ever see my memories, not even if they open up my skull and poke around in my brain. You won't be able to find *Ghostbusters* in there (even though I could recite the entire screenplay). It's not stored as a distinct unit in my brain; rather, it's a set of connections between cells.

But not all the connections that make our memories are the same. They can be stronger or weaker—and the memories more vivid or vague, respectively—depending on how the connections were forged and how often they are reinforced. Whether we like it or not, some of the worst things that have happened to us are also some of the most memorable. This is partly because a burst of stress hormones is a memory booster (while long-term stress is not). It's almost like our body is saying, "Wow, this is super terrible. Let's remember this so that we can hopefully avoid it in the future." And things you practice often, like singing a song every day, can become robust memories over time.

Now, setting aside the instances when you forgot someone's name or couldn't remember the bonus question on a quiz, your memory right now is storing an impressive amount of information you can pull up at will or with very little effort at all. You have incredible data storage. Your memory is why you can even make sense of this book, since you've committed the meaning of these words to memory. Your memory is why you can remember the people in your life and navigate your social world. And your memory is why you probably get through an average day without hurting yourself—you remember that the stove burner is hot, that knives are sharp, and that pulling a cat's tail will result in carnage.

But we can (and usually do) put too much stock in our memories. Memory is magnificent, and our repeated survival is a testament to its effectiveness, but it's equally interesting to think about how our memories can be very, very wrong.

There's never a guarantee that our recollections are accurate. Our memories store information that we can later retrieve, but the system is vulnerable to distortions. Like a document stored on a computer, a memory can be edited each time you access it. Indeed, every time you retell a story, your brain has the opportunity to overwrite the older version with what you just said. So in a way, you are playing a game of telephone with yourself every time you recount something. And just like telephone, it's possible to play it and come out with most of the original meaning intact, but it can also veer off into sheer absurdity.

It's startlingly easy to misremember something and be quite sure you are right. I know I have had childhood memories muddled with those of my siblings. I was convinced that my brother had once been sleepwalking and shoved his comforter into a small cabinet. But years later when I said, "Ha, remember when you did that totally weird thing?" he looked at me and said, "Dude, that was *you.*" Upon further reflection and fact-checking with various family members, I updated that memory file, admitting to myself that clearly that particular memory was wrong.

Now, maybe I can be forgiven because I was sleepwalking, after all, so the key part of the story happened while I wasn't fully conscious, but it was unsettling to think that I had filed that story away with someone else at its center. How often has this happened with no nearby fact-checkers handy to set me straight?

I pride myself on generally having a good memory (the sleepwalking misfile notwithstanding), which is a dangerous game to play. While I do remember many fine details of conversations (or movies I've seen) and easily memorize factoids and numbers, I might give my memory more credit than it is due. When I read the latest neuroscience news about how unreliable our memory is, I think about the decades-old grudges I still hold or embarrassing blunders I revisit in the wee hours of the morning when sleep evades me. What if they weren't as bad as I remember? Is my skewed memory doing me a great disservice? What if over the years of rehashing the original incidents, I have made them worse than they truly were? It's an unset-

tling feeling. I can't simply unmoor all my memories and pretend the past doesn't matter. It seems the most scientific thing to do, bereft of any photographic or written evidence for certain bad memories, is to *remember* that they might not be as bad as I think, and get on with my day.

While it's distressing to reflect on the limitations of our memories, since so much of our identities are wrapped up in them, just remember that we're all in the same boat. So go easy on yourself, and forgive that person who forgot your name that one time.

Personalities and Clashes: If Only Everyone Was *Just Like You*

One of the weirder things about us humans is how different our brains can be. They all work essentially the same way, with neurons sending chemicals that relay messages. They have the same general areas: the motor cortex that controls some of our muscles; the prefrontal cortex that governs our behavior and fun things like planning ahead; the hippocampus that consolidates memories; and a whole bunch of

other stuff we don't have time to get into. And yet two people can look at the same painting, read the same book, or hear the same news story and have wildly different takeaways. And somehow simultaneously, everyone also thinks they alone are right about everything, and that the people who disagree with them are stupid, crazy, or controlled by injected government nanobots.

This is often frustrating, especially when we're trying to build consensus and move forward with a group decision, but I also find it inspiring that people can be so different and have disparate viewpoints. Diversity is a good thing, especially when it comes to ice cream flavors, movie casting, and our brains.

When you think about it, it can be advantageous that there would be so much variability. This randomness is partly to thank for our current status. If people weren't so different, maybe no one would have ever thought to build a fire, fashion a wheel, use a pulley, or make a pizza.

Just mathematically, if people interpret and respond to things in lots of different ways, it increases the chances that someone, somewhere will get it right. And this is one of the many reasons diversity in the various fields of human endeavor matters so very much.

Take science for example. Some people wonder why something with the reputation for being so objective would need to have diversity in order to be successful. But they're only thinking about part of the picture. Sure, in science you carry out tests and experiments to get results. It shouldn't matter too much who is carrying out an individual trial. The results ought to be identical; the same way that any two people following a cake recipe should get equally good cakes if they carefully measure and adhere to the procedure. But science is not merely the following of directions.

Science is mostly about asking questions and figuring out ways to test them. That's where you need lots of different viewpoints, ideas, and experiences—that is, different brains. Different people will think of different questions entirely. And people with different experiences might think of different ways to test them. In all areas of our human existence, whether it's government, education, industry, anything at all, really, a lot of different ideas and insight is better than a single story.

Sometimes I think the world would be a better place if people *weren't* so different, and we agreed far more often. But let's extrapolate this to an absurd level. What if everyone in the world was just like you. Imagine it: if we were all interchangeable, with the same likes, dislikes, and priorities. It might be nice if all restaurants served the food you like, and every book was something that interests you. But then everything would start to be predictable and boring. No third-act film twist would ever startle you. Plus, in this hypothetical reality, we would also all share the same fears, blind spots, and prejudices. In a universe of Beatrices, no one would write novels, go to outdoor music festivals, or open microbreweries. But I'm thrilled that other people do those things.

Have you ever seen something and thought, "Huh, that's clever. I never would have thought to do that"—like bacon-wrapped dates or a blanket with sleeves? That, right there, is why it's good that people are so different, and not everyone is just like you (even though I'm sure you're wonderful). Diversity matters. No single person, no single type of person, can do everything or even conceive of everything. We all have something to offer.

Stressed Out: How Our Brains Are Just Trying to Protect Us

Our body's stress system is about as annoying as the people who stress us out, which I can say with authority because I'm stressed out and annoyed with people so often. It's frustrating, upsetting, and sometimes nauseating. But just like those seemingly annoying people, with a different perspective, our stress system isn't so bad.

Think of something that stresses you out. Public speaking, heights, going to a big party where you only know the host. Perhaps your heart starts to race. You might

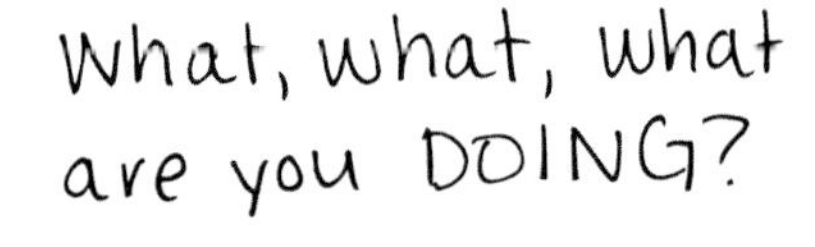

feel your mouth get a little dry. You get some butterflies in your stomach. This happens to me all the time, and I often beg and plead for my body to knock it off and let me move forward with an awkward encounter without the added feelings of dizziness and sweaty armpits.

What sometimes helps is to remember that my body is trying to save me; it just doesn't always work very well. My body's stress system has not adapted to my twenty-first century lifestyle. It's still looking out for me as if I live in the wild, interpreting mild social stress like a life-or-death encounter with a hungry lion.

Stress causes some real changes in our bodies. When you perceive a threat, whatever that might mean to you, your body makes hormones that travel through your bloodstream to put your whole body on high alert. It's like we have a switch inside of us that flips from "relax" to "totally freak out."

When that stress switch is turned on, the heart starts to pump faster, since we need our muscles to have plenty of oxygen, obviously, seeing as how we're about to run a mile away from whatever the trouble is. Wait—we do need to run, don't we? Oh, we don't. Hm. Well, the heart will keep chugging away just on the off chance we do. Also, seeing as how we are all about short-term survival and not long-term health, let's stop digesting our food. We don't need to be thinking about breaking down pasta right now; we could be dead in five minutes. Hold off on that until the threat has passed. Let's also up the sweat production because when the water in the sweat evaporates, it will cool us down, and like I said, since we're about to sprint away from danger, we want all our marathon systems up and running.

Yeah, thanks so much for doing this to us, body. If it happened once a month, it would be fine, but this sort of stress, day in and day out, can take quite a toll on us. However, while this is far from a pleasant experience, I'm going to go on record saying that I'm grateful even for my overly active stress system. It reminds me to take care of myself. If I didn't have it, who knows where I'd be by now. Something has to tell me to slow down, sit back, and have a cup of tea. Because I probably would never do that otherwise. I'd be working myself to death and feeling guilty every time I stopped.

Maybe prolonged stress is like accrued vacation days. It's a reminder of the time you should take off. I've heard of a newfangled notion at some tech companies that they will stop counting vacation days, in theory making vacation days unlimited, so people can take more time off. But you know what could also happen? Without a number of days a person has earned, they might take *less* vacation. I know I would have a hard time putting in a request for two weeks off at such a company if I didn't have the displayed numerical value of days off I had most undoubtedly accrued there to back me up.

So when you feel stressed, and your body shows any of the symptoms of chronic stress—headache, fatigue, constipation, insomnia, any of these super fun things—remember that it's your body, in its own very overprotective, obnoxious way, trying to take care of you, and it means it's time to take a mental health day, stay in bed, and read a good book or three.

Caring for Our Blobs: A Moment to Ponder Mental Health

Speaking of mental health and self-care, let's get into some of the wonderful ways we can nurture the wrinkly mass of neurons in our skulls. Sometimes when I'm thinking about my own brain this much, I find myself placing my hands on the top of my head and thinking, "Yup, it's in there." It's protected in a bath of fluid encased inside my reasonably strong skull, but what a fragile machine the brain is. All of this—my memories, identity, sensory abilities, and all of human civilization before and after me—is possible because of nothing more than a three-pound blob of soft, puckered pink tissue.

Because our brains are so complex, there are lots of ways for things to go wrong. Brain injuries are one thing. Without enough oxygen, cells can die; a blow to the head can disrupt brain functions; and a seizure or a stroke can damage some of the circuitry. But aside from those injurious issues, the brain can also sometimes have trouble without sustaining a physical trauma. Yes, I'm talking about the many mental illnesses and maladies the human brain is prone to.

Can you tell which brain belongs to someone with depression?

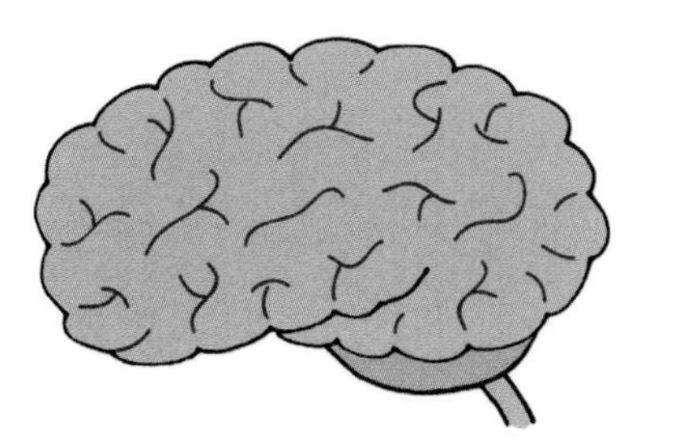

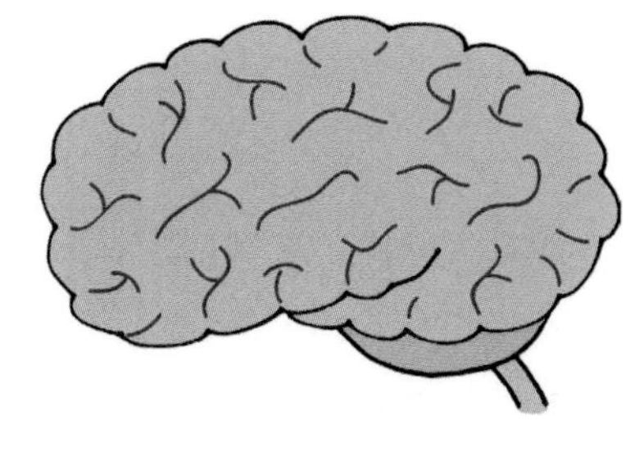

ME NEITHER!

Perhaps nothing is so common and yet so stigmatized as mental illness. There are hundreds of diagnosable mental disorders, some of them familiar ones like depression and anxiety. Millions of us have them, but despite how common mental illness is, it is treated like something exotic or shameful that should be overcome with willpower, a burden we never bestow on people with other chronic conditions like rheumatoid arthritis or type 1 diabetes.

Don't take *medication* for your rheumatoid arthritis! Just ignore it and it will go away on its own.

To lay my brain's cards out on the table, I have in fact been diagnosed with one of these disorders, depression (even though my perception of this condition is usually runaway, overwhelming anxiety).

But I'm lucky to live in a time when I have options. I can make an appointment with someone whose entire job is helping people think about their brains. That has not been the case throughout history. If I'd been born a hundred years ago, I shudder to think what would have become of me if I had experienced the same symptoms. Today, this specially trained person I meet with can assist me in unraveling the mental knots in my head, learning techniques like meditation, and pondering how my other habits like diet, exercise, and sleep are affecting my head jelly. What a time to be alive.

And on top of that we have a suite of medications available to people suffering with anything from bipolar disorder to anxiety. Which means that people whose brains make too much or too little of certain neurotransmitters (chemicals our brains use to send signals between cells) can take supplements to adjust their brain chemistry so they can succeed. Who says that every chemical in your body has to be homemade? Taking a daily medication isn't so different from needing a cup of coffee to dust away the mental cobwebs at the start of the day.

But no matter the state of your mental health, don't take your mind for granted; think about your thinker. How this folded, squishy organ works is absolutely mind-boggling, and it's wonderful that we can do things to help it. So, please, do what's best for you, and let people do what's best for them. And if anyone judges you for taking care of your mental health, let me know, and I'll smack them upside the head.

CHAPTER 9

A Big Family Tree: Genetics and Evolution, and How All Living Things Are Related

Within each of your thirty trillion cells (well, most of them), you have long strands of deoxyribonucleic acid, or DNA. This spindly molecule contains the instructions for building all the tiny pieces that make you (1) a living thing, (2) an animal, (3) a member of the species we call *Homo sapiens*, and (4) your unique self.

This molecule is the way life on Earth has come to store information and pass it through generations. Even though DNA is relatively simple as far as structure goes, it has somehow managed to perform this function for billions of years. It's a molecule that you may have heard is shaped like a twisted ladder. The sides are alternating groups of sugars and phosphates, and the rungs of the ladder are pairs of bases, of which there are only four sorts—adenine, thymine, cytosine, and guanine.

Every living thing on this planet, regardless of exactly what kind of creature it is, has DNA. It's one of just a handful of things all organisms on Earth have in common. Whether it's an animal, plant, fungus, bacterium, archaeon—it doesn't matter. We all have DNA. We speak the same cellular language.

This molecule that determines part of our identity also ties us to our many relatives. We all have DNA because we are all related. It's not an exaggeration, metaphor, or hippie-dippie daydream. And when I say "we," I don't just mean *Homo sapiens*. We're related to bananas, mushrooms, and bacteria too. We share a common ancestor, even if that ancestor was a great-great-great-great-great-a-whole-bunch-more-greats-that-I'm-too-lazy-to-type grandmother. And the DNA in your very cells right this minute contains a trail of helical breadcrumbs we can use to trace our evolutionary histories and common ancestry, whether you believe it or not. So let's talk about your DNA, where you got it, and how it links you to every living thing on Earth.

Your Very Own DNA: Your Unique Sequence

Somehow you came to pick up this book and decide of your own free will to keep turning its pages. Countless decisions, big life ones and small daily ones, led to this moment, but partly you are reading this book today because of your DNA. So if you're not having fun, you know who to blame.

Your DNA makes you special and your particular brand of *you*. Even though there are now and have been in the past so many billions of people on Earth, it's a near mathematical certainty that no other person has had your exact sequence of DNA. In at least one small way, we're each unique.

Your genome, the entire set of your DNA, is three billion letters long. Those three billion rungs in the DNA ladder make up your own personal instruction manual, and almost all of your cells have a copy (most notably, red blood cells do not). Thirty trillion cells each with three billion letters—these numbers are too big for me to comprehend. You have so much DNA in your body, I don't know how you have room for anything else. But it's not just you. I have three billion letters too. That's part of what we have in common, fellow human being. Members of the same species have remarkably similar genomes, which is part of how we determine what species are these days.

This molecule, DNA, has done a lot for you. It contains the information that, during development, ensured you turned into a person and not (as I would have maybe preferred) a koala. It has instructions for making proteins that determine your hair color, blood type, and whether or not you can curl your tongue. Some of your DNA controls other parts of your genome, determining where genes are used and where they are ignored. Genes for making nails, for example, are active in your fingers and toes, but not so much in your liver (thank goodness). There are sections of our DNA whose purpose remains a complete mystery. They may be spacers or

structural support for other genes, like paragraph marks in a Word document, or they could have functions we haven't discovered yet. And who knows—your DNA could hold the key to these future scientific discoveries.

If you're of the rebellious sort, you probably bristle at the thought that a molecule inside your body has so much control over your life. I know I do. So let me mention that DNA isn't everything. Our experiences, everything from the food we eat (or don't eat) to the books we read and people we know, play a large part as well.

If you cloned yourself, that individual with initially the same DNA sequence as you, exposed to alternate experiences, might choose a different career, play a different musical instrument, or prefer different pizza toppings than you. Or maybe they'd think and act like you down to their favorite brand of shoes. It's hard to tell. When identical twins are separated at birth and adopted by different families, sometimes they lead startlingly similar lives, and other times they don't. How rigid our identities turn out to be is yet another thing partly determined by the symphony of our many interconnected genes.

If you spend too much time thinking about DNA in general, it's easy to forget that it's in your body—in each of the cells that keep being you, minute after minute, day after day. This very instant, while you scan over this page and make meaning from these symbols, some of your cells are making copies of their DNA, going through rung by rung and making a whole new set. This way, when the cell divides, it can divvy out a complete copy of DNA to each daughter cell, the two cells resulting from cell division. And other cells are cracking open the DNA instruction manual to find the recipe for making the lactase enzyme, an estrogen receptor, or a component of hemoglobin, whatever the case may be.

Now, I don't know much about you, but (unless you're a clone) I can almost guarantee you got your DNA from two individuals, which we often call a mother and a father. And where did they get it from? They each have their own two individuals who gave it to them, who in turn got it from two more. On and on this list goes. Because of our species' generation length and life span, we usually think mostly about our parents and grandparents. Sometimes we get to know our great-grandparents too (my great-grandmother died at the age of 101 when I was in college). But we don't think much about our great-greats, much less our great-great-great-grandparents. Not only did they die long before we were born, but there are increasingly too many peple to keep track of, as the number of individuals doubles with each generation. I know the names of my eight great-grandparents, but if you held me at knifepoint I couldn't tell you anything about my sixteen great-great-grandparents, and so you can definitely forget about my thirty-two great-great-great-grandparents. They are quite the enigma.

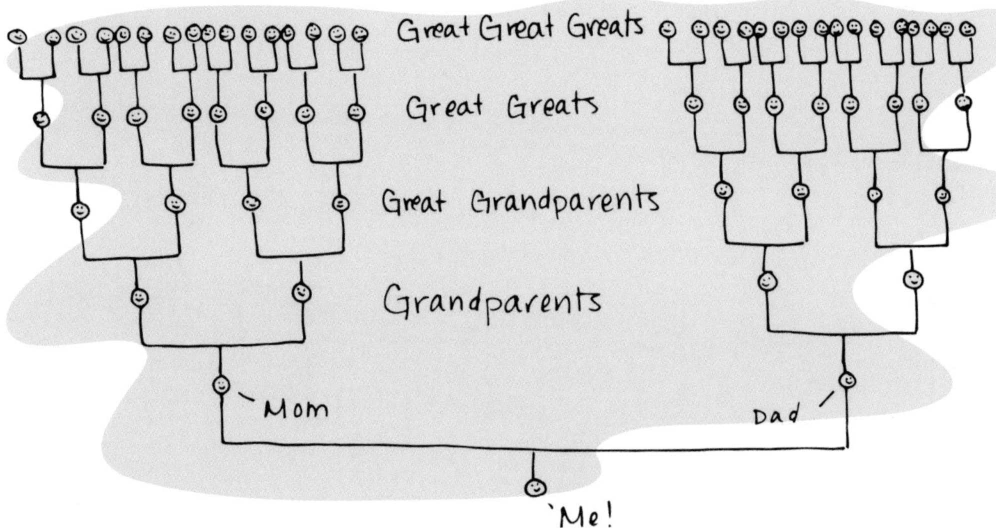

This exponential growth of ancestors means that a mere ten generations ago, there were over one thousand people on Earth who could claim me as a direct descendent. You may also have noticed that this rate of ancestry is unsustainable. If you keep multiplying, around thirty-five generations back you get to a figure that exceeds the number of people who have ever lived. This is easy to reconcile: some of your family tree's branches overlap due to inbreeding.

But inbreeding aside, when you think about your many, many ancestors with direct ties to you, you can get a better idea of how incredibly dispersed our heritage is. We humans put so much stock in our ancestry, but beyond a certain point, it's nearly meaningless because of the sheer number of previous people we can claim in our past.

Through the National Geographic Genographic Project, my DNA can reveal which historical figures I'm somewhat related to. The results were a tad underwhelming. Apparently Nicolaus Copernicus (the mathematician who realized that the sun is the center of our solar system) and I share a great^x-grandmother who lived during the Ice Age (more than twelve thousand years ago). But that far back, I simply have so many grandmas, sharing one with Copernicus (even though I think he's super) isn't all that impressive to me. Apparently, I also share Ice Age grandmas with Queen Victoria, Marie Antoinette (and her mom, Maria Theresa), Napoleon, and Benjamin Franklin. They're my very (make that extremely) distant cousins. (But then, so are most people of Western European ancestry.)

No matter how many grandmas you share with Napoleon, you got the sequence in each of your cells from a long line of ancestors. Somehow you are unique and yet completely derivative. Your recycled, repurposed genome, along with your choices, has made you who you are today. And if you want to, you can pass the genetic baton to a new human or gaggle of humans. This molecule that tethers us to the past also holds the sequence we can send into the future. It's the closest we'll probably ever get to time travel technology.

Neanderthals Among Us: The Evolution of Our Species

After thinking about your many ancestors who have lived in the last few thousand years, it's time to go even farther back and think about your ancestors hundreds of thousands or even a million or two years ago. Going this far back, the family tree starts to get pretty blurry. We can't tell you for sure what these ancestors from long ago looked like exactly, and we're still mapping out all these lineages and timelines.

Our species, our ancestors, and our extinct relatives are very hard to define. I mean, many organisms are challenging to classify—we don't even have snails figured out yet—but humans are especially tricky because we keep moving the goal posts, not least because we keep on moving. Sometimes what makes two species different by our understanding is simply that they live in different places and never come into contact. Asian elephants and African elephants fit that mold. And on the flip side, species that seem to be similar to us, that all live in the same area, can in fact be different species that don't interbreed, like giraffes. We thought that there was only one species of giraffe, but it turns out there are four distinct ones, and we just didn't notice.

Humans throughout history have made things especially difficult because they moved around and crossed what otherwise could arguably be species barriers, having babies with what would otherwise be easily separated human species.

You see, today we people are all one species, *Homo sapiens*, but thousands of years ago there were more species of humans that lived at the same time, sometimes alongside one another. The question is, where do we draw the line between entirely different species of early humans, and just different communities of early humans? That's not up to me to decide, and I'm grateful for that. Human ancestry is ridiculously complex. But that's what makes it fun to think about over lunch.

Because of that widely used, simplistic sketch that shows human evolution happening in a linear way in about five steps following an apelike ancestor gradually standing upright, our ancestry is as misunderstood as it is complicated. Yes, over time people have changed, and if you go back in time, you'll come to a point when there were ancestors of ours who looked a great deal like today's apes. But evolution doesn't happen by cloning with upgrades the way that diagram looks. You don't have one person making the next, more advanced person. Evolution happens with whole populations and the slow changes over time in the genetics of entire communities. It doesn't show the family tree of our species. In fact, even a family tree doesn't quite cut it here because that's too neat and tidy for the garbled chaos that is the business of life and ancestry.

The human family "tree" is more like a wild, unpruned, sprawling family ivy. Some vines branch off in different directions, some run into an obstacle and stop, but they all emanate from the same root system if you dig down deep enough.

But even this updated botanical analogy fails us because sometimes distinct branches partly or completely merge back together. Neanderthals are a great example. For a while they were a human evolution punch line. With their pronounced brow line and chunky bones, they were the epitome of the simple cave man (although I've always wondered why we never speak similarly of cave women, cave tweens, and cave toddlers). But the more we learn about this "extinct" group of early humans we realize they were pretty handy, and also pretty *friendly* with *Homo sapiens*. Many people of European ancestry appear to have some Neanderthal DNA sequences lurking in their genome, meaning that *Homo sapiens* and *Homo neanderthalensis* were making babies together. We say that Neanderthals went extinct, but pieces of them live on in people like myself, so even the concept of "extinction" isn't cut-and-dried.

And while Neanderthals are the most well-known relative of our species, there were others, like the Denisovans who lived in Asia around forty thousand years ago. The same way that Europeans may have traces of Neanderthal DNA, some communities in Asia can find Denisovan DNA sequences in their genomes.

All this means that if we hopped into our time machine, we don't know how far back we'd have to go to find people so vastly different that we couldn't assimilate into their world and become part of their community, making babies and all. But if you did make it back and blend with some Neanderthals, you might get pretty bored between bouts of foraging for food and hunting woolly rhinos. I'd recommend bringing this book with you to pass the time.

Our Cousin, the Banana: How All Living Things Are Related

While our DNA has souvenirs from our recent past, it also has mementos dating back much farther, as in hundreds of millions of years. We can find these when we compare our genomes with nonhuman species. When we do this, we can see there is a certain amount of overlap. For example, we supposedly share about 60 percent of our genome with bananas. And 90 percent with cats. But obviously we are not more than half banana. Or mostly cat.

First of all, some of those statistics can come from looking not at the entire genome but just genes that code for proteins. And when it comes to making proteins, no matter what sort of living thing you are, you need to do a lot of the same basic stuff, often with the same suite of cellular machinery and proteins. We all need to get nourishment from some sort of food, we all have to get rid of wastes, and we all have to replicate our DNA so our cells can split in two and keep living. You know, the basic housekeeping of life. Bananas, cats, humans—when you get down to the world of the cell, we have a lot in common, so it would make sense that we share a lot of this everyday DNA. It's just like a car; no matter how basic or fancy your vehicle is, it still probably has seats, a steering wheel, and windows. Some features are unavoidable.

But bananas, cats, and humans didn't develop all their DNA sequences independently and happen to arrive at the same answer. We share so much DNA because we have a common ancestor from a long, long time ago. The genes in our bodies that code for hemoglobin (the molecule in our red blood cells that carries oxygen) are strikingly similar to genes that code for hemoglobin in plants (yes, some plants have hemoglobin). That's not a coincidence, and it's not because hemoglobin is the only possible way to transport oxygen. It's because we share an ancestor that had a form of those genes, and we're all still playing from the same deck of cards.

And even though some of these more shocking stats (like sharing so much of our genes with a banana) are a tad misleading, I like the perspective they allow. The things that make us different are actually quite minor, and it all depends on your frame of reference. To someone who studies bacteria, all animals, from sponges to elephants, are pretty much the same. But people who study great apes probably spend a lot of time thinking about the few differences that lie between us and bonobos.

Since we share so many traits with bananas and cats (and other living things, but let's stick with these two examples for now), you'd think that among our fellow humans, we'd see we have so much in common it's almost laughable to emphasize any differences amongst ourselves. Unfortunately for us, our favorite pastime is finding differences. This has served us well in many respects, as it's the backbone of the whole scientific method—trying different things and noting what changes between trials, as well as observing the natural world. But then we do that to ourselves and force a wedge between us and the very people we have so much in common with that it's nearly absurd.

What we really need is a vastly different being that will make all our squabbles seem insignificant. I'm holding out hope that hostile aliens will show up in warships any day now. It's our only shot at world peace; after the aliens' defeat, that is. (Although, knowing us, it would take only about two generations to forget all the lessons we'd learned.)

I find little else so humbling and comforting than remembering that I'm a relative of amoebas, dinosaurs, and snails. So is your coworker who sends passive-aggressive emails and the impatient person behind you in line at your local coffee shop. We are made of the same stuff. We're connected through time and space. With just a few chance encounters and rolls of the genetic dice did we humans emerge with big brains and social skills, solving problems and sometimes even working together to be able to ponder big ideas and learn how much we are related to other living things. It could very easily not have happened. There was never a guarantee that an animal here on Earth would evolve the intelligence to drastically change its surroundings and build fires, structures, and spacecraft that leave the solar system. If we could turn back time and run this Earth simulation again, we might just as likely not appear.

We like to think of ourselves as the pinnacle of evolution, or we think that we are somehow "more evolved" than other living things, but that's not true. All living things here today have been evolving for the same amount of time. Sure, some creatures have stayed noticeably the same for many years. The horseshoe crabs on the beach today bear a remarkable resemblance to fossils we find that are four hundred million years old. But they're not "less" evolved than we are. They simply stumbled across a great gig early in the game. Other lineages changed, found new niches, and plenty (in fact, the overwhelming majority) went extinct.

Every living thing you see (or don't see) around you, from the bacteria on the backs of your hands to your pet pooch and the cherry tree down the block, are your long-lost relatives. I would say to treat them as such, but I know that most of us don't send birthday cards to our second cousins once removed, so instead, think of them as friends.

What We Did to Dogs: All about Artificial Selection

One of the most surprising things about traits controlled by DNA is the relative ease with which we can alter them. Just look at dogs. I mean, seriously, I need you to locate the nearest dog immediately and gawk at it. What is it—a spaniel, a terrier, a shepherd? No matter what it is, that sort of creature didn't exist anywhere from a few hundred to a few thousand years ago. First of all, wow. But second of all, what is wrong with us? Why did we do this?

It may have started innocently enough: *These two dogs are good at doing some sort of thing. I bet their puppies would also be good at the things.* Once people realized that after many generations of careful breeding they could cultivate specific traits in dogs, that was it. Hair color, hair length, size, general temperament, snout shape. We molded and changed it all. Things quickly got out of hand, and now we have bulldogs that can barely breathe.

To say that something is genetically modified might engender gasps and glares depending on your audience, but the truth is that we have been tinkering with DNA long before we had any clue what this molecule even was, much less how slight mutations in DNA between individuals can cause natural variation. Dogs are evidence of this, but so is just about all the food we eat.

Consider corn. If you got into that time machine yet again (although we might need to bring this in and get it serviced after so much use) and went back nine thousand years, you wouldn't be able to find a recognizable cob of corn anywhere in the world. This mainstay of backyard barbecues, not to mention a key ingredient in many tasty chips and junk food, did not exist. Instead, you'd find the naturally occurring ancestor of today's corn. The cob was a fraction of the size and had only a few kernels, which were covered in strong casings that took some serious smashing to get through. Over the course of hundreds to thousands of years, people in Central America selectively bred these corn ancestors, transforming this average grassy

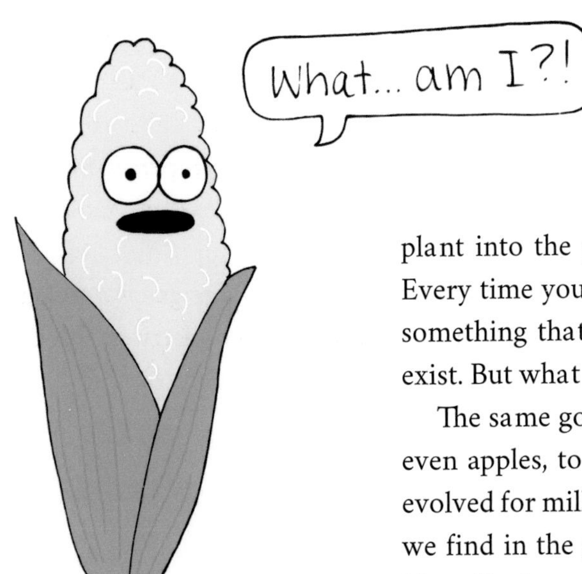

plant into the productive food crop it is today. Every time you have a corn chip, you are eating something that you could argue shouldn't even exist. But what a delightful abomination it is.

The same goes for peaches, watermelons, and even apples, to name but a few. The plants that evolved for millions of years didn't produce what we find in the grocery store today. We did that. Now, I'm just as interested as anyone in delicious heirloom tomatoes and other forgotten crops, but we're kidding ourselves if we think that a varietal that's a hundred years old is somehow "natural" (whatever that even means any more). Most of our food is as altered as the Chihuahua.

So, again, while genetic engineering is a touchy subject these days, we've been doing this for ages by choosing which plants, dogs, cows, and chickens breed and which don't. And even amongst ourselves, when we choose a mate, we're in one small way potentially electing the genetic future of humanity.

So stare at a dog while you eat corn on the cob and contemplate how neither of those things would exist if we hadn't taught that old DNA some new tricks.

Cellular Stowaways: Viral DNA and Mitochondria Are Evolutionary Souvenirs

We've talked a lot about your particular DNA, but I should also mention that some of it isn't "yours." What I mean is, besides the DNA that shapes who you are, you have stowaway DNA that belongs to something else—tiny creatures that live inside your cells. They're called mitochondria, and they're pretty weird.

Our mitochondria are a relic from a chance encounter hundreds of millions of years ago. As best we understand it, these components of our cells used to be free-living organisms that struck a deal with a larger cell, trading their tendency to produce helpful energy-storing molecules for the luxury of an indoor lifestyle. They're sort of semidomesticated, like house cats. They're there for us, but they also have their own agenda.

These mitochondria have their very own DNA that doesn't mix with ours. They set their personal schedule of division, though they still get portioned out into the two daughter cells that form when our cells divide. And their DNA can be very handy for establishing lineages since it doesn't get reshuffled with half from each parent like the rest of our DNA does. Nope, our mitochondria are an as-is, wholesale purchase. You get them just from your mom, as only the egg contributes them to the single cell that becomes a person (or cat, or guinea pig, or whatever).

They're inside you right this instant, toiling away to help your cells, while maintaining their own identities. They're a vital part of you, but they're also not really, truly you. You can ride this identity crisis for the rest of the day, or you can be thankful for these cellular tenants and all they do for you. The choice is all yours, unlike your DNA.

And as many of my favorite infomercials would say, *but wait, there's more.* It's not only mitochondrial DNA we have hiding out in our bodies. We also have DNA from viruses past.

Part of what makes viruses so tricky to deal with is the way they go about making more of themselves. As something that is not quite a living thing, a virus reproduces by hijacking other cells and forcing them to do its bidding. Part of this entails injecting cells with viral genes. Sometimes the virus takes over our cells, making copies of its viral genome and destroying everything in the process. But other times the virus inserts its genes into our DNA and doesn't do much else. And occasionally that DNA can get handed down to the next generation.

As much as 8 percent of the human genome came from viruses. Although it sounds very body-snatch-y, this viral DNA is part of what makes us who we are. Some viral genes are most active when we are but a tiny fetus in our mother's uterus, which might mean they are crucial for our development. These viruses could have been a factor in our evolution, as well as other creatures' evolutionary trajectories. We likely wouldn't be where we are today without these viruses. And as an added bonus, these viral genes can help us trace ancestries across the tree of life. Some viral DNA is shared by all mammals; some is shared by mammals and fish. It can help us establish relationships, by pinpointing a place in time when we all shared a common ancestor who got infected with some virus.

It's another way that our DNA grounds us in history. We got it from our parents, who got it from their parents, who got it from their parents, who got it from their parents, all the way back to the beginning of life on Earth. You and every single organism alive on the planet today descended from an uninterrupted chain of ancestors that reproduced successfully. And the things alive today, every creature from bacteria to us humans, are just the tiniest percentage of life that has ever existed. But parts of those long-dead organisms are still with us. And hundreds of years from now, some small part of you will still exist in future living things. Even if you don't have children, or your children don't have children, you are still part of a big group of related organisms, and they will live on. (Probably.)

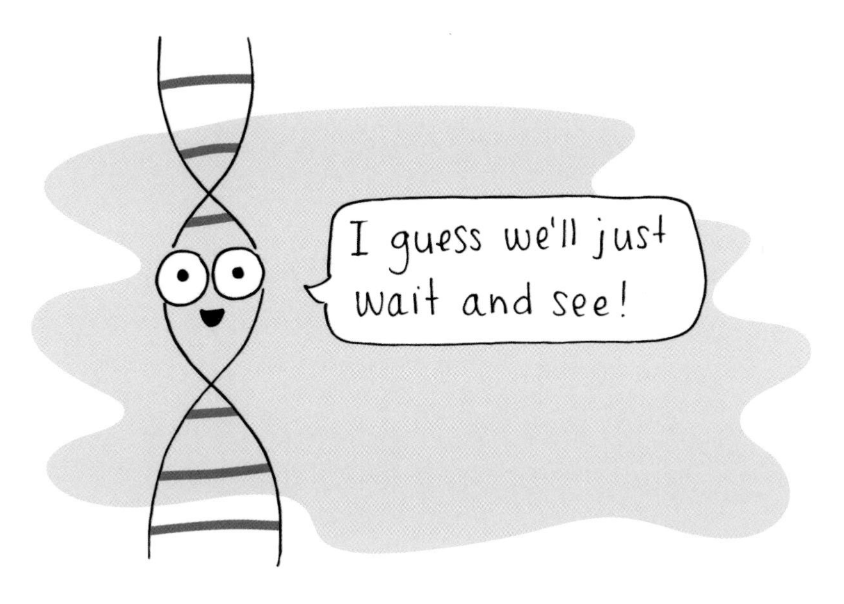

CHAPTER 10

Poop Is Relative: Nothing Really Goes to Waste, Including *You* When You Die

Poop is relative. To me, it's highly disgusting, but to a bacterium, it's a delicious, fortuitous feast. It's easy to discount poop. I mean, it sure doesn't look like much, but like so many of the often unexamined things we do on a daily basis, our wastes are truly fascinating (not to mention humbling).

It's something all living things have to contend with. We need a way to jettison stuff. For single cells, it's not so hard: just push junk outside of the cell membrane, and you're done. But for big animals like us, some more infrastructure is required. Somehow over the eons of evolution, the system we settled on was poop and pee. We excrete a lot this way. Over the course of your lifetime, you take in and leak out so many atoms you're scarcely the same person you were when you started.

The atoms you have now won't be yours forever (if they were ever even "yours" to begin with), and not only because some of your individual cells die and you excrete atoms daily. Someday you will put all your atoms up for sale; you will pee and poop no more. You will be dead. But that's not such a bad thing. So let's talk about poop, death, and decay. It'll be fun; I promise.

The Scoop on Poop: The Breakdown of an Average Poop

Life isn't always dependable. In fact, sometimes it can seem downright chaotic. But one thing is always true. We all have to poop—even (or especially) when we don't want to, like when we're killing time at the airport, attending a party, or sharing a hotel room with a friend.

But hey, at least poop is consistent. As an anxious person, I'm all about things I can depend on. Yes, no matter how flaky your friends, unpredictable the world

news, or turbulent your family life, at least you (and everyone else) is an invariable pooper. And if you don't feel productive today, at least you can count the feces you're making right now. It's not glamorous, but it's something.

Now, let's get down to business, starting at the beginning, where all good stories and poops ought to start. Every day you put some food in your mouth hole, and then you probably forget about it. I mean, I do. But it's in there, worming its way through your gastrointestinal tract and, while you're enjoying your afternoon, your body is trying its hardest to extract as much nutrition from your food as possible, which is probably an especially tall order if you ate junk food for lunch like I did.

We talk about our digestion (or indigestion) often enough, but what does it mean to digest something anyway? Let's wade into some gastric juices for a little while and think about this. To digest something means to disrupt it or partially destroy it and loot it of anything useful. It's very cutthroat. Our bodies are trying to break our food into tiny, tiny pieces. Our mouths do this physically, by stomping all over the food and making it mushy, but along the rest of its route, it's beaten up chemically, by acids and enzymes that attack its chemical bonds.

Right this minute, the sauces in your stomach are hurling acid bombs at your food, trying to disrupt its very nature, breaking bonds inside those big molecules to cut them into smaller and smaller chunks so they can cross over into your bloodstream and be disseminated throughout the body. It's nothing personal; it's simply a logistical consideration. You have a lot of hungry cells that need to get fed. Your body is like an expert caterer that knows how to feed thirty trillion individuals each and every day without fail.

Yes indeed, day after day, your body uses chemistry to beat the crap out of your breakfast. And it's not just acid at work. After the acid bath in your stomach, your meal moves into your small intestine and gets a squirt of bile, which was made by your liver. Bile is alkaline, as in the opposite of acidic. Bile does a lot: it neutralizes some of the acid that the food got from the stomach, it helps break down fats, and it gets rid of wastes at the same time. Color me impressed.

Your liver is constantly filtering things from your blood and adding them to bile. One of those wastes is bilirubin, which is made when broken-down old red blood cells are discarded. Bilirubin is a substance you are very familiar with, as it's what makes your poop such a lovely shade of brown.

But the liver isn't just responsible for dyeing your poop; it also breaks down all sorts of molecules in your body, including drugs and toxins. When you take a painkiller like acetaminophen (Tylenol), the reason it wears off is mostly because your liver grabs it out of your blood and disposes of it. This is something I like to remind people when they have lingering concerns about general "toxins" in their bodies, or they want to go on "cleanses." You have a liver. That's part of its job. The reason that Tylenol doesn't relieve pain for the rest of your life (as cool as that sounds), and the reason the effects of caffeine (or alcohol) wear off after a few hours, is because your liver (and your kidneys too) is running a constant cleanup system. All you have to do is keep drinking water (and not overdo it on

the Tylenol—or anything else). So if by going on a cleanse you mean you'll be drinking sufficient water, then please proceed.

But let's get back to the stinky stuff happening in your intestines. While your digestive system wages war on the molecules you've eaten earlier today, there is one in particular that doesn't flinch at the acid bombs, enzymes, and other digestive tactics our bodies use. That would be cellulose, which is made by plants. Cellulose is a particularly sturdy molecule that gives plants their structure, making stems rigid and woods woody. We call it fiber. And although it's counterintuitive, it's important for a portion of our diet to be this completely indigestible molecule.

The weirdest thing about fiber is that it looks very similar to starch. Both of these are rows upon rows of linked glucose molecules. (Glucose as in sugar.) Starch is linked with a bond that our bodies readily break, so we access all those glucose molecules, which have plentiful energy stored in their bonds, making starch a calorie-rich molecule that countless diets instruct you to avoid at all costs. "Step away from the potatoes!" these diets tell you. But I never listen to this carb-ist stance. Some of my best friends are potatoes.

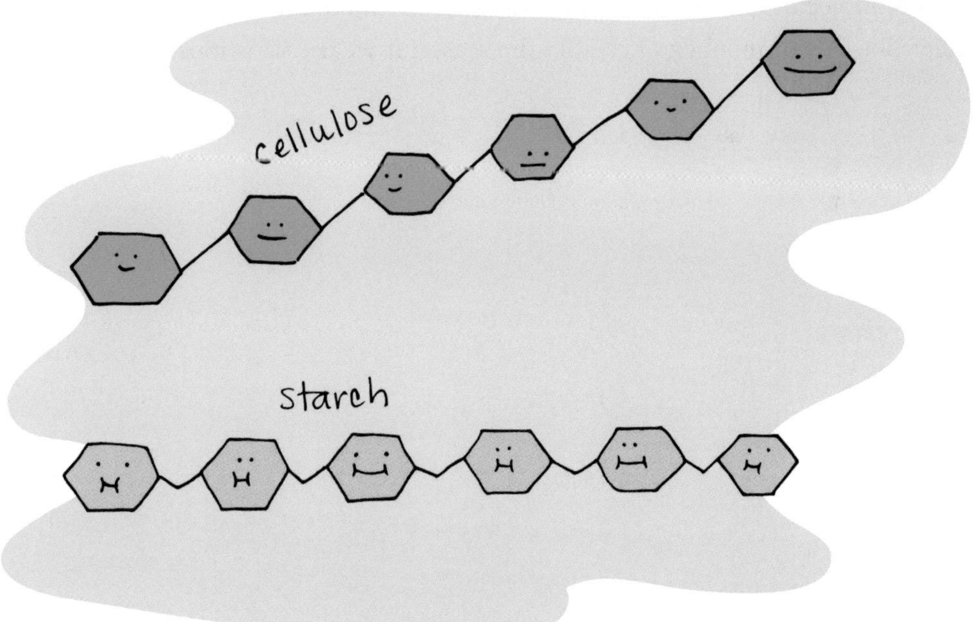

Cellulose is also made of chains of glucose, but these chains are linked in such a way that we can do nothing with them. Our digestive tract hurls everything we've got their way: we chew them up, we physically break them down, and our stomach lobs acids at them. The cellulose cares nothing for these threats. It moves through unaltered. This brave molecule is an important part of our diet because it keeps the system moving forward, shepherding material through your intestines that might otherwise meander. And in the end it makes your feces softer and bulkier, so it holds together and is easier to, um, push out.

The biggest contribution to your poop is the unsung hero of digestion: water. The precursor poop in your intestines is like a thick river flowing through your body, and water helps move things along just like fiber does. We even have a word, *constipation,* for what happens if your stool has too little water and becomes a challenge to expel,

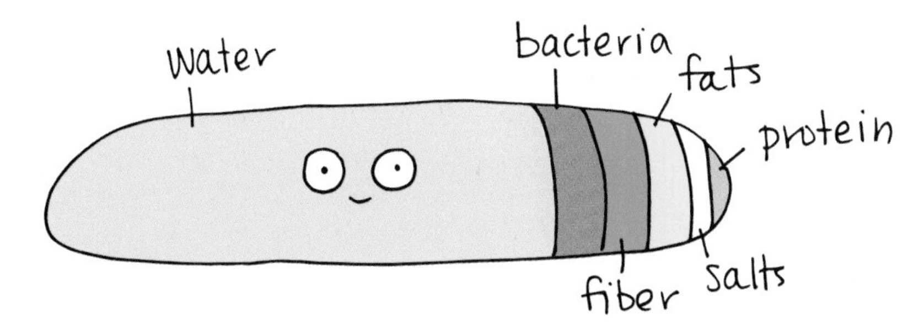

But poop isn't only the remains of our meals after we've physically and chemically accosted them. Poop is so much more. Our excrement also contains living things, such as the bacteria that reside in our guts. Think about how they must feel about that. These bacteria were perfectly happy living in our digestive tract, but every day some of them get swept up by the poop stream, and they are forced to exit the body. They'll probably get over it, though. At least they're not going through this experience alone.

We'd probably think more highly of poop if it didn't smell so bad. The odor is largely because of the bacteria we have in our guts. When the bacteria eat our food, enjoying it just as much as we did when we ate it, they poop too. It's like double poop going on. And their waste products are smelly compounds with sulfur, which smells like rotten eggs. But ponder this. There is nothing truly, inherently foul about those substances. Our brain detects them and deciphers that they smell bad. Other creatures could find them neutral or even think they smell divine. But *we* think they smell awful because of associations and some genetically inherited wiring. It's to our benefit that we perceive these things as noxious. Imagine what would happen if someone thought poop smelled *good*. They wouldn't be disgusted by it, and might think nothing of keeping it around or (shudder) eating it.

Some other planet:

But then, it's actually pretty weird that something that normally inhabits one part of our body, such as our colon, can have disastrous effects if it finds its way into another part of our body, let's say our eyes or mouth. The body has very clear lines drawn, and it's important that everyone stay in their lane for things to go smoothly.

So before you flush, take a moment to applaud the work your body did to transform pizza and ice cream into a water-bacteria-cellulose sausage. Poop is a great equalizer. No matter who you are, how rich, glamorous, or powerful you might be—we all have to sit on a porcelain open seat and push out yesterday's food. And no matter what, it stinks all the same. No amount of money will change that. Oh, and that old tip to picture people in their underwear if you're nervous in front of them—forget that. Picture them on the toilet. That's way better.

You and Urine: Why We Need to Pee

I think it's annoying that we have two forms of excretion: poop and pee. I have to urinate multiple times per day, and it's inconvenient to do so while I'm out running errands. Can't we combine these two things the way that birds do? They only have one exit. Or at the very least, can't we get this all over with once a day? The answer is, evidently, no. We mammals have decided to have two ways to excrete waste. And okay fine, we'll keep this setup. It has been working pretty well, I guess.

All this peeing has to happen because your body is constantly running a filtration system to keep it nice and clean. It pulls wastes into one area and then dilutes them with water so it can wash them out. Just like we have to clean our house (which I probably don't do enough), our bodies are constantly having to tidy themselves. And luckily, my body is more committed to cleaning than I am.

With pee, it's all about the kidneys. They're constantly filtering your blood, taking water, salts, sugars, and other small molecules out, as well as some things broken down by your liver. But the kidneys don't remove everything in some kind of indiscriminate cleaning rampage. They keep things in balance, maintaining optimal levels of these molecules in your blood supply.

One of the important things we get rid of in our urine, and what we named it after, is something called urea. When our cells break down proteins, the process results in some waste products that have nitrogen in them, and nitrogen is a tricky thing to get rid of because it can turn into ammonia (NH_3) in our cells. This substance, the very same molecule as the stinky cleaning product, is dangerous, and your body doesn't want a bunch of it floating around in your blood.

Rather than contending with troublesome ammonia, our cells spend precious energy and resources to turn it into something called urea. They take two molecules of ammonia and shackle them together along with a carbon atom and an oxygen atom. This chemical isn't hazardous, so our cells shuttle it around without worry. It's a bit like chemically handcuffing the ammonia so it can't attack anyone on its way out. Urea can safely travel through our bloodstream and get pulled out by our kidneys.

The kidneys also have to constantly take into account how much water you have available. The blood has a range. If you guzzle down a big glass of water, more of it will be in your blood, so your kidneys will pull that out, and add it to your pee. But if you've been exercising and haven't had enough water, your kidneys keep as much water in your blood as possible. But when I say it like that, it sounds weird. How do the kidneys know? Well, they don't. The same way a water molecule doesn't know where it is. The kidneys manage to do their job based on a series of switches that regulate this entire filtration system. The process of making pee is a chemical dance. A lot of it goes back to water's qualities, and how things dissolve in it.

Let's think specifically about how your kidneys "know" when you're getting dehydrated. This touches on how just about every chemical function in your body works. We have all these means to tell the concentrations of things in our body. It's all based on chance encounters between molecules. When water dips low, that means the concentration of salts and other things dissolved in the water becomes higher. Those salts interact with triggers in your body that notice that there is more salt than there was before. Imagine that I tasked you with estimating the number of people in a banquet hall, but stipulated that you accomplish this with a blindfold and earplugs in place. All you could do is bumble around the room and tell me how many people were there based on how many you bumped into. That's what it's like to be a molecule in a cell in charge of gauging the concentration of something.

But anyway, chance encounters consistent with dehydration cause hormones like one called the antidiuretic hormone to be released. These hormones then go to the kidneys and start a chain of events that eventually causes these organs to conserve more water and make more-concentrated pee than they would otherwise.

It's all a numbers game: chemicals moving around, chance encounters, and probabilities. This is the fundamental way our body works; in fact, it's how all living things work. No single protein or molecule "knows" what it needs to do, but with the systems our cells set up, billions of chance encounters wind up making everything work so you can go on living, and also peeing.

Feces As Food: Not All Living Things Think Poop Is Gross

Wastes are surprisingly fascinating, as are the systems in our bodies that produce them. But it's also important to stay away from them. Those of us with functioning toilets luckily don't have to spend much time thinking about this, but a water shut-off or plumbing issue quickly reminds us of just how troublesome our wastes are.

Yes, poop is our daily biohazard—your everyday brush with something that could make you very sick if you let it. Even a healthy human poop has bacteria like *Salmonella* and *E. coli*, which can give you food poisoning, especially if you introduce them to your dinner before you eat it. But poop can also spread diseases like hepatitis A (caused by a virus); diarrheal illnesses like cholera (caused by bacteria); and intestinal parasites like *Giardia* (caused by a protist, a diverse group of mostly single-celled critters that include things like *Paramecium* and amoebas).

And yet, not everyone thinks poop is so bad. Decomposers positively adore the stuff. This is the lovely name we give to any living thing that likes eating wastes. But again, they don't think of it as waste. To decomposers, it's merely another meal.

Bacteria, some insects, and fungi are the main players of the decomposition world. We rely heavily on them to process all the death and waste that we and other living things make. They recycle it and change all this garbage into usable material.

If I were to, uh, poop outside and monitor the progress of this solid waste over time, I would be able to see it slowly disappear as the atoms in my poop would get reshuffled by these many decomposers. But I swear to you, I haven't done this and will not ever.

But my dog, who is not a human biohazard, poops outside in my backyard. I usually pick it up, but sometimes a poop stays out there for a while (don't judge me; I've got other crap to take care of), and it slowly disappears. The water in the poop evaporates, possibly floating up into the sky and joining a cloud that could rain down on people hundreds of miles away.

And bacteria and fungi munch on the carbohydrates and proteins in the poop as well. Some of the atoms from those molecules are assembled into carbon dioxide, which could float to a nearby plant that could use it to build a ring of glucose, which could then be eaten by a leaf-munching insect. (Yes, I do think this much about my dog's feces.)

On a different scale and timeline, your poop meets similar ends. Enormous waste treatment plants depend on tiny living things that eat our poop, transforming it into different compounds that can get reused. The atoms in poop don't stay there for long.

In the end, just like one woman's trash is another woman's treasure, waste is in the eye of the beholder (but that literal situation can cause pink eye, so wash your hands). And we should be very grateful for the decomposers that love our waste so much. If they didn't constantly clean up after us, we'd be in some pretty deep...you know what.

When *We* Are Waste: Someday You'll Die, but That's Okay

You should appreciate every one of your bowel movements because someday you will poop for the last time. All poops are a countdown to this final fecal moment. Our excrement is a daily reminder that someday we, too, shall share the same fate.

But wait, have you reflected upon your mortality yet today, or is this the first session? I'll admit I'm among the most morbid of people. I'm not a card-carrying death

expert or anything, but I do dabble in death pondering now and then. I find it quite inspiring to think about my limited time in this existence and enjoy considering what I would like to do with it. But it still doesn't stop me from wasting countless hours rewatching seasons one through three of *Arrested Development*.

When I imagine my inevitable death, I usually picture myself sitting in a cozy chair at the age of ninety-five or so, muttering to no one in particular about what a good movie *Pacific Rim* was, suffering a heart attack, and dying (hopefully quickly). My last words will probably be something like, "Corpse poo." However, I could easily die today in a car crash, or tomorrow from bad bagged salad. Or, and very realistically, any day now my cells could go rogue and form a cancerous tumor that could spread throughout my body and shut everything down. There are so many options when it comes to death.

But however I go, I will one day undoubtedly do so. And for the record, I'm not going to go into the whole transhumanist thing—you know, the future where people upload their consciousness into some virtual reality where they could "live" on forever or something. Or a sort of *Futurama*-like existence where people who should be dead are instead immortal heads in floating jars. That sounds absolutely awful to me. No, thank you. It's good that people die. It makes space available for the next generation and all their ideas. Don't be selfish. Give up the table so other people can have dinner at the fancy restaurant.

Right now, though, for a variety of reasons, I consider myself very much alive. My heart is pumping blood around my body, I'm sucking air into my lungs in a regular fashion, my liver is breaking down toxins, my kidneys are filtering my blood. Meanwhile, cells in my brain are chattering amongst themselves so I have the wherewithal to write this sentence. That's sort of the standard version of being alive. I can say with a great deal of confidence that you fall into this camp too.

But someday one of those systems will fail. Eventually all of them will, and then I will die. Let's assume just for fun that my body is relatively intact, as opposed to being eaten by a school of fish or burned to an absolute crisp. (Comforting thoughts, both of them.) So now I'm dead, and I have a very good-looking corpse left behind. What now? Skirting around the entire funeral industry, let's say I decompose the old-fashioned way. My individual cells, starved of oxygen, die and fall apart. My body starts to liquefy.

What gives? Where's the oxygen delivery?

Apparently the heart stopped. We have a few moments before we die.

Ohhh... gotcha. Well, been nice knowing you.

The pleasure's all mine.

Part of the process of decaying involves the bacteria that have been on and inside me the entire time. Like starving housecats, they start to eat me, since they're not being fed on a regular basis anymore. Without immune cells to protect me, and with vulnerable cells everywhere you look, the buffet is open.

In our current culture, we don't want anyone to eat us, even if we're dead. We go to extreme lengths to keep this from happening, circulating chemicals into our bodies that decomposers hate, or really snubbing the decomposers by burning the body and transforming it into ashes.

But when I die I'd like to be buried as naturally as possible. I truly don't mind the thought of being eaten by bacteria and other decomposers. I want to go the way of all the poops I've ever pooped, being recycled and turned into new things. My dying desire will be for the atoms I've been borrowing from the universe to shuffle on with their existence as quickly as possible. I guess I'm atomically impatient.

Element Recycling: What Happens to Your Atoms in the End

The coolest thing about death is that, as my body starts to become much simpler molecules like carbon dioxide, water, methane, and so on, I can supply atoms and molecules for other living things to use. My carbon could be used by plants. My water could support a colony of fungus. The calcium from my bones could be used by a snail to make its shell, or it could get incorporated into a rock in Earth's crust and stay that way for millions of years. All of my atoms will wind up somewhere.

Of course, if I am buried in a very substantial coffin, I will hang on to those atoms for a pretty long time. But like I said, I don't plan to hoard them. And even if you are into atom-collecting (which is fine, seriously, no judgment—if you need more time with them, I understand), nothing is eternal. Even if you are in the most tightly locked coffin of all time, the land you're buried in will change eventually. It could plunge into a sinkhole. Oil deposits could take it over. If you're buried near a fault, an earthquake could split it in half. In a few billion years it could even wind up getting

pushed back down underneath another crustal plate, becoming metamorphic rock like granite (with a fossil of you in it maybe).

And okay fine, let's say you manage to hold on to a lot of your atoms for the next five billion or so years, when our sun will expand into a red giant and fry our planet in the process. Then you will, most surely, get recycled, becoming cosmic dust swirling in space and contributing to a new solar system. Maybe some of your atoms will wind up in a sun, a comet, or a planet. Maybe the new planet will be able to support life, and a few of your atoms will be used in an alien with a body plan we can't even conceive of with our Earthly imaginations. But you will get redistributed eventually. You can't hold on to those atoms forever, no matter what you do.

That is seriously what I just said.

The continuity of matter is what drives our entire universe. Nothing goes away. Things only move around and change. When the world seems hopeless and your life feels like it's getting off track, this can be a comforting thought. Things may change, but nothing can ever truly be destroyed.

We have this time on our planet, a blip in the long history of Earth, which won't be here forever. Make the best of your life while you have your atoms on loan from the universe. That's all any of us can do. And while you're at it, appreciate the mostly empty atoms around you, whether they're crystals cooked up by our planet, plants that make the food you eat, or arranged to spell out the DNA sequence in your body that has been handed down to you from a line of organisms dating back billions of years. And be nice to your fellow moving, breathing, thinking cellular poop factories. We're all just trying to stay properly hydrated as we get through another rotation of Earth, bathed in the radiation from our sun, receiving radio waves from the dawn of the universe, surrounded by dark matter we can't find.

Isn't it amazing?

References and Further Reading

Chalabi, Mona. "What Are the Demographics of Heaven?" *FiveThirtyEight*, October 14, 2015. https://fivethirtyeight.com/features/what-are-the-demographics-of-heaven/

Eschner, Kat. "There Are Four Giraffe Species—Not Just One." Smithsonian.com, September 12, 2016. www.smithsonianmag.com/smart-news/there-are-four-giraffe-species-not-just-one-180960411

Friedland, Andrew, Relyea, Rick, and Courard-Hauri, David. *Environmental Science: Foundations and Applications*. W.H. Freeman and Company: 2012.

Gazzaniga, Michael S., Ivry, Richard B., and Mangun, George R. *Cognitive Neuroscience: The Biology of the Mind, 4th Edition*. W.W. Norton & Company: 2014.

Grotzinger, John and Jordan, Thomas H. *Understanding Earth, 7th Edition*. W.H. Freeman and Company: 2014.

Human Origins Initiative. "What Does It Mean to Be Human?" The Smithsonian Institution's National Museum of Natural History. http://humanorigins.si.edu

Reece, Jane B., Urry, Lisa A., Cain, Michael L., Wasserman, Steven A., Minorsky, Peter V., and Jackson, Robert B. *Campbell Biology, 10th Edition*. Pearson: 2013.

Tro, Nivaldo J. *Chemistry: A Molecular Approach, 4th Edition*. Pearson: 2017.

Wolfson, Richard. *Essential University Physics: Volume 1, 3rd Edition*. Pearson: 2015.

Yong, Ed. *I Contain Multitudes: The Microbes Within Us and a Grander View of Life*. HarperCollins: 2016.

Zimmer, Carl. "Ancient Viruses Are Buried in Your DNA." *The New York Times*, October 4, 2017. www.nytimes.com/2017/10/04/science/ancient-viruses-dna-genome.html

Index

C

D

E

F

G

H

I

S

T

U

V

W

Z